Elasto-Plastic Damage Behaviour of Concrete Elements

Elasto-Plastic Damage Behaviour of Concrete Elements presents the results of practical experiments with numerical analyses and case studies, along with a summary of basic theory, to provide an accessible explanation for young practising engineers on the performance evaluation of concrete structures.

It shows how the mechanical phenomena of familiar concrete structures can be expressed using mathematical models and provides a solid basic understanding of the nonlinear behaviour of concrete structures. It applies elasto-plastic theory to damage mechanics and the modelling of cracks in concrete, drawing on the author's 25 years of design and construction experience as a professional engineer, as well as recent research.

- Sets out the reality of damage mechanics in concrete
- Connects standard theory with good design and construction practice

The book is suitable for structural design engineers and researchers.

Elasto-Plastic Damage Behaviour of Concrete Elements

Hidenori Tanaka

CRC Press
Taylor & Francis Group
Boca Raton London New York

CRC Press is an imprint of the
Taylor & Francis Group, an **informa** business

First edition published 2023
by CRC Press
6000 Broken Sound Parkway NW, Suite 300, Boca Raton, FL 33487-2742

and by CRC Press
4 Park Square, Milton Park, Abingdon, Oxon, OX14 4RN
CRC Press is an imprint of Taylor & Francis Group, LLC

© 2023 Hidenori Tanaka

Reasonable efforts have been made to publish reliable data and information, but the author and publisher cannot assume responsibility for the validity of all materials or the consequences of their use. The authors and publishers have attempted to trace the copyright holders of all material reproduced in this publication and apologize to copyright holders if permission to publish in this form has not been obtained. If any copyright material has not been acknowledged, please write and let us know so we may rectify in any future reprint.

Except as permitted under US Copyright Law, no part of this book may be reprinted, reproduced, transmitted, or utilized in any form by any electronic, mechanical, or other means, now known or hereafter invented, including photocopying, microfilming, and recording, or in any information storage or retrieval system, without written permission from the publishers.

For permission to photocopy or use material electronically from this work, access www.copyright.com or contact the Copyright Clearance Center, Inc. (CCC), 222 Rosewood Drive, Danvers, MA 01923, 978-750-8400. For works that are not available on CCC please contact mpkbookspermissions@tandf.co.uk

Trademark notice: Product or corporate names may be trademarks or registered trademarks and are used only for identification and explanation without intent to infringe.

Library of Congress Cataloging-in-Publication Data

Names: Tanaka, Hidenori, author.
Title: Elasto-plastic damage behaviour of concrete elements / Hidenori Tanaka.
Description: First edition. | Boca Raton : CRC Press, [2023] | Includes bibliographical references and index.
Identifiers: LCCN 2022048434 | ISBN 9781032256160 (hbk) | ISBN 9781032256177 (pbk) | ISBN 9781003284246 (ebk)
Subjects: LCSH: Concrete--Deterioration. | Concrete--Plastic properties. | Concrete construction--Case studies. | Fracture mechanics--Mathematics.
Classification: LCC TA440 .T226 2023 | DDC 624.1/834--dc23/eng/20221207
LC record available at https://lccn.loc.gov/2022048434

ISBN: 978-1-032-25616-0 (hbk)
ISBN: 978-1-032-25617-7 (pbk)
ISBN: 978-1-003-28424-6 (ebk)

DOI: 10.1201/9781003284246

Typeset in Sabon
by Deanta Global Publishing Services, Chennai, India

Contents

Preface ix

1 Introduction 1

    1.1 Background 1
    1.2 Overview 3

2 Theory of elasto-plastic damage mechanics 7

    2.1 Background 7
    2.2 Mechanical characteristics of concrete 7
    2.3 Concept of continuum damage mechanics 10
    2.4 Elasto-plastic damage constitutive laws 13

3 Strength characteristics of concrete by elasto-plastic damage mechanics 17

    3.1 Background 17
    3.2 Static compressive strength 18
        3.2.1 Experimental results 18
        3.2.2 Stress-strain curve for compressive side 19
    3.3 Static tensile strength 22
        3.3.1 Experimental results 22
        3.3.2 Stress-strain curve for tensile side 23
    3.4 Static bending strength 26
        3.4.1 Experimental results 26
        3.4.2 Stress release of crack elements 27
        3.4.3 Analysis of static bending strength 29
    3.5 Bending fatigue strength 31
        3.5.1 Experimental results 32
        3.5.2 Bending fatigue strength analyses 35

4 Structural experiments and numerical analyses 45

4.1 Background 45
4.2 Bending fracture of reinforced concrete cantilever beams with carbon fibre sheets 45
    4.2.1 Characteristics of carbon fibre sheets as reinforcing materials 46
    4.2.2 Experimental results 47
        (i) Bending fracture by monotonous loading 47
        (ii) Bending fatigue by cyclic loading 48
        (iii) Bending fracture after pre-fatigues 51
        (iv) Pull-off test 52
    4.2.3 Numerical analysis by the finite element method (FEM) 55
        (i) Identification of stress-and-strain relationships by material tests 55
        (ii) Bending fracture analysis of RC beam reinforced with carbon fibre sheet 55
        (iii) Bending fracture analysis of RC beam with carbon fibre sheet by pre-fatigues 57
4.3 Adhesion fracture analyses of carbon fibre sheet reinforced concrete 60
    4.3.1 Adhesive fracture analysis by monotonous loading 61
        (i) Preliminary analysis 61
        (ii) Adhesive fracture analysis by the improved equivalent stress 62
    4.3.2 Adhesive fracture analysis by cyclic loading 64
4.4 Accumulative damage of fibre sheet caused by negative thermal expansion coefficient under cyclic temperature 68
    4.4.1 Cyclic temperature change tests 69
    4.4.2 Relationship between temperature and strains for specimens 71
4.5 Adhesive characteristics between inorganic injection materials and concrete in the post-installed anchor method 73
    4.5.1 Influence of injectable diameter and length 75
    4.5.2 Improvement of adhesive force by the wedge-shape effect 76
        (i) Adhesion characteristics of injection material and concrete 76
        (ii) Adhesion characteristics of anchor and injection material and concrete 78

    4.5.3 Numerical analyses by return mapping algorithm 80
 4.6 Cumulative damage to interfaces due to repeated temperature loads of two inorganic materials with different thermal expansion coefficients 81
    4.6.1 Experimental method 82
    4.6.2 Experimental results 83
    4.6.3 Effects of roughness 85
 4.7 Shear strength of concrete 86
    4.7.1 Experiments 86
    4.7.2 Analyses 86

5 Applicability of damage mechanics to the concrete field 89

 5.1 Background 89
 5.2 Applicability of damage mechanics 90
    5.2.1 Representative volume elements that define material strength and constitutive laws 90
    5.2.2 Structural members that define structural performance 91
    5.2.3 Time to define performance maintenance 92

Reference books   95
Index   101

Preface

Concrete technology was introduced roughly 140 years ago and has been used as a construction material to build a large amount of social infrastructure that supports economic activity.

During this period, remarkable technological progress has also stimulated economic activity. On the other hand, it is also true that natural disasters such as earthquakes, typhoons, and floods have caused the loss of lives and property.

I think one of the ultimate goals of engineers and researchers is to predict more accurately material failure and structural collapse and minimize damage caused by natural disasters. Of course, the subject should also extend to the deterioration of structural performance due to deterioration over time.

For this reason, new knowledge has been obtained from continuous experiments and numerical analyses, and it has been reflected in design standards and guidelines from time to time. In particular, numerical analysis in structural design has been able to obtain remarkable results and knowledge through the spread of the finite element method. In addition, an analysis method for handling discontinuous surfaces has been proposed, including a finite cover method that can evaluate the progress of cracks.

In order to evaluate the mechanical properties of concrete structures, including changes over time, it is important to understand the occurrence and evolution of cracks.

In this book, I introduce examples of our efforts, mainly in finite element analysis and experiments based on continuum damage mechanics that can evaluate structural characteristics, including the occurrence and evolution of cracks in concrete structural elements. The introduction of basic theories has been simplified, and as much space as possible has been devoted to the introduction of experiments and numerical analysis examples at the practical level.

I hope that it will help engineers and researchers in charge of structural design.

Lastly, I would like to express my gratitude for the support I have received from so many people. In particular, I would like to express my deepest gratitude to Yutaka Toi, Professor Emeritus of the University of Tokyo, for his guidance on damage mechanics, and to Mr Tony Moore, CRC Press of the Taylor & Francis Group, for giving me the opportunity to write.

Chapter 1

Introduction

1.1 BACKGROUND

Japan adopted reinforced concrete bridges in the early 1900s, and over the past 100 years or so, concrete has been used as a construction material that is indispensable for the formation of social infrastructure such as bridges, dams, tunnels, and port facilities. During this period, the principles and technical concepts common to each structure were established as standard specifications, and since the first edition in 1931, they have been revised or revised in accordance with academic and technological advances. In particular, Japan is one of the major earthquake-prone countries, and experienced enormous damage in the southern Hyogo Prefecture earthquake, which triggered a growing momentum for safety performance inspections of social infrastructure, and materials for earthquake-resistant technology were developed separately. The accumulation of these findings is largely due not only to experimental approaches, but also to the improvement of analysis techniques.

The application of analytical technology to concrete structures increased dramatically with the spread of the finite element method in the 1980s and the improvement of the performance of electronic computers. In particular, general-purpose analysis codes such as NASTRAN, ADINA, and ABAQUS have become indispensable in the design field because they can evaluate the behaviour of complex structures that could not be evaluated by each technical document that targeted relatively simple structures. However, at that time, evaluation up to a region close to elastic behaviour was limited, and there were relatively few examples of research on nonlinear behaviour as a concrete structural member.

Thereafter, using fracture mechanics targeting fracture phenomena caused by the occurrence and growth of cracks, tensile softening characteristics were incorporated into numerical analysis and theory, and the fracture phenomenon of concrete due to the cracking progress was examined, and the smeared and discrete crack model were proposed; and not only the tensile failure of concrete, but also shear failure and compression failure. It has been used to study the nonlinear behaviour of reinforced concrete members. Although these models can be highly effective in evaluating the softening phenomenon of the concrete and the load-bearing capacity of the member, there are issues such as when tension, compression, bending, deformation, etc., are mixed, and there is dependence on the mesh in the direction of crack development, etc., which suggests that it is a somewhat unsuitable method for practical design.

On the other hand, continuum damage mechanics (hereinafter referred to as damage mechanics) is a theoretical system constructed by Kachanov, Lemaitre, Murakami, and others since the 1960s. The feature of this mechanics is that structural deterioration such as a decrease in stiffness, toughness, etc., due to the generation and growth of microscopic voids such as micro-cracks or micro-voids inside the material, and a decrease in the remaining life can be expressed within the framework of continuous mechanics. In addition, since damage mechanics is based on a continuum, it is considered that combination with the finite element method can be performed relatively easily.

However, at present, damage mechanics is not necessarily fully utilized. This is due to the fact that it is not used in general-purpose codes, that the comparison with experiments is insufficient, and that there are few examples of verification of constitutive laws.

Regarding the application of damage mechanics to concrete, there are relatively many examples of research at the level of concrete materials, among which there are constitutive laws using multi-phase models by Suaris et al. and models that extend this to high-strength concrete. However, although structural members such as beams are sometimes seen as targets, they express the phenomenon of the deterioration of bending rigidity and structural bearing capacity well and can be considered as an effective design tool for safety performance evaluation. It should also be noted that in Japan there are relatively few examples of the application of damage mechanics to concrete structural elements.

In recent years, regarding the design of concrete structural members, a method of evaluating lifecycle costs that takes into account the deterioration of structural functions due to deterioration over time has been adopted, and maintenance management has been shifted to a method of planning from the design stage. This deterioration over time indicates a decrease in durability and load-bearing performance due to fatigue by repeated loads, salt damage, carbonation, and other chemical reactions, and quantitative evaluation of this phenomenon has become an important technical issue in order to extend the life of social infrastructure and perform efficient maintenance.

Damage mechanics was first proposed by Kachanov to evaluate creep rupture and has since been applied to fatigue failure life prediction and creep fatigue failure evaluation, mainly for metallic materials. This is because by considering the amount called the damage variable, the cumulative damage degree of materials and structural members can be understood, and the presence or absence of the occurrence of cracks can be expressed relatively easily in relation to this damage degree and the critical value.

Thus, while damage mechanics is mainly used in the field of metals, it is a natural progression to apply this mechanics to maintenance techniques such as predicting the remaining life of concrete structural members. In particular, regarding the safety of concrete structural members, since it is required to continuously satisfy their performance, it is necessary to verify the transition in performance after repair and reinforcement performed for the purpose of restoring function.

From this perspective, this book applies damage mechanics to the evaluation of the reinforcing effect of carbon fibre sheets on concrete structural elements, fatigue life, and adhesion characteristics of heterogeneous materials and verifies their effectiveness by comparing them with experiments.

1.2 OVERVIEW

Rational maintenance of existing concrete structural members will become an indispensable technology in the future. However, to quantitatively understand the change in the concrete structure over time, it is necessary to continuously measure the state amount such as the stress-strain relationship of each constituent material and the displacement amount of the structural member over at least 50–70 years, corresponding to the service life as a guide, which is basically difficult. In addition, reproducing the deterioration state in experiments involves many uncertainties and preconditions, so it is evaluated within limited environmental conditions, and there are physical limitations.

In addition, technical documentation on the maintenance of concrete structures in Japan has been prepared since 2000 but is insufficient.

In recent years, technology for improving the strength of existing concrete structural members using carbon fibre sheets has attracted attention. This technology is a construction method in which carbon fibre is arranged in one direction, resin is impregnated to process into a sheet shape, and epoxy resin or the like is attached to the concrete surface layer as an adhesive to increase member strength. However, the design materials of this method are mainly compiled based on experiments, and the scope of application is relatively limited. The features of carbon fibre sheets, such as their light weight (about one-fifth that of steel), high strength (about ten times that of steel), high elasticity (about the same level as steel), and corrosion resistance performance are not sufficiently utilized. In addition, since this sheet reinforcement is relatively easy to construct in the field, the demand has more than quadrupled since the 1995 earthquake in southern Hyogo

Prefecture, and it is widely applied to pillars, beams, plate structures, etc. Considering the development potential of this method, it is desirable to establish an evaluation method for the strength, durability, or life prediction of concrete structural members reinforced with carbon fibre sheets.

In addition, the anchor method, in which mechanical equipment such as blowers are fixed to concrete structures, is used in various fields including civil engineering and architecture, and issues related to their durability have been pointed out. For the problem of the deterioration of adhesion strength between concrete and heterogeneous materials, it is important to understand the adhesion characteristics between steel anchors, injection materials, and concrete and to evaluate the fracture modes.

Therefore, this book summarizes the mechanical behaviour of concrete structural elements reinforced with carbon fibre sheets and the adhesion characteristics between concrete and heterogeneous materials from both experiments and numerical analysis by damage mechanics.

The following is an overview of each chapter.

Chapter 1 briefly describes the background and reasons for the review of concrete structure analysis methods and the reasons for applying damage mechanics to concrete, and describes the purpose and outline of the book.

Chapter 2 briefly describes the mechanical properties of concrete and explains the basic concepts of damage mechanics, which is the theoretical subject of the chapter. Furthermore, the process leading to the elasto-plastic damage constitutive laws based on the potential theory is specifically described.

In Chapter 3, based on the results obtained in the strength tests, the material constants in the constitutive laws in contrast to the elasto-plastic damage laws described in Chapter 2 are identified. At this time, stress release expressing the critical value of the damage variable and the brittle behaviour after cracking is incorporated into the finite element method and analyses.

The target strengths of concrete are compressive, tensile, bending strength, and bending fatigue strength. When evaluating the number of load-fatigue failures at bending fatigue strength, Fully Coupled Analysis, which analyses all models, and Locally Coupled Analysis, which targets local elements and integration points, are performed, and differences and characteristics between the two are clearly indicated.

In Chapter 4, the elasto-plastic damage constitutive laws identified in Chapter 3 are incorporated into the finite element method: The mechanical properties of the concrete structural members, specifically the effect of improving the bending bearing capacity of a beam reinforced with carbon fibre sheets; the reduction of strength due to the repeated loading of the reinforcing beam; the adhesion fatigue of the carbon fibre sheet and the concrete block; the cumulative adhesion damage due to the repetitive temperature load between the carbon fibre and the concrete block with a negative thermal expansion coefficient. The results of verifying the adhesion strength between the steel anchor, the injection material, and the concrete, and the interfacial peeling between heterogeneous materials due to the presence or absence of surface treatment, from both experimental and analytical perspectives, are presented.

Finally, Chapter 5 describes the applicability of damage mechanics to concrete structures from three perspectives: Representative volume elements that define the strength of materials, structural members that define structural performance such as beam columns, and time that defines the deterioration of durability and load resistance with ageing.

Chapter 2

Theory of elasto-plastic damage mechanics

2.1 BACKGROUND

The basis of structural design is to understand the relationship between cross-sectional force and strength resistance caused by external loads. These are performed by structural analyses, which are generally conducted using the finite element method.

Therefore, in this chapter, when modelling as structural elements, I describe the mechanical characteristics of concrete centring on the constitutive laws, which are the basic information necessary for analysis. Particularly as concrete is often used as reinforced concrete members in combination with reinforcing bars, and since the reinforced concrete structures allow cracking and there is an interaction between the reinforced bar and the concrete accordingly, the laws as a material are not sufficiently reflected in the structural design.

In the current design, the calculation formulas such as strength other than bending resistance, for example, shear strength, and torsional strength resistance are empirical formulas, and those are evaluated independently of the constitutive laws. Based on this situation, this is a feature of concrete, focusing on continuum damage mechanics that can express crack progress in concrete structures.

Regarding the application of continuum damage mechanics, which is the main subject of this chapter, to concrete structures, I introduce the outline of damage variables and effective stresses, which are the quantities of mechanics uniquely handled by this mechanics, in physical representation. In addition, the strain equivalence principle used in the process of deriving the constitutive equation is schematically shown, and the elasto-plastic damage constitutive equation incorporated into finite element analysis is described in detail.

2.2 MECHANICAL CHARACTERISTICS OF CONCRETE

The compressive strength of concrete is about a dozen times greater than the tensile strength, and its mechanical behaviour is considered as a

DOI: 10.1201/9781003284246-2

representative brittle material. Therefore, concrete alone is limited to gravity dams and mountain tunnels. In general, concrete is used as reinforced concrete in combination with reinforcing bars placed on the side where the tensile stress occurs and is often handled from a macroscopic point of view, and the constitutive laws are also considered according to the stress state as a member. Since the tensile strength of concrete is extremely low compared to compression, the laws of the conventional design code are not described on the tensile side but are limited to the compression side.

The laws of concrete are considered to be uniaxial, biaxial, and triaxial as stress states, and are affected by strain rate, temperature, etc. Uniaxial and biaxial stress states generally occur in concrete structures. Examples of biaxial states include two-dimensional structures such as wall structures. The triaxial state is assumed for triaxial compression, biaxial compression, and uniaxial tensile, the former occurs in columns and beam members surrounded by shear reinforcing bars, etc., and the latter occurs when compressive stresses and bending stresses are simultaneously applied.

In the case of a structure in which compressive stress is generated in any cross section such as a special member that mechanically introduces compressive stress in the axial direction such as a prestressed concrete, the constitutive laws under triaxial stress can be applied, but the restraint effect that can be considered by the weight of the superstructure, like a column, may not be expected depending on the load conditions, and careful consideration is required for application.

On the other hand, it is considered that the laws of concrete in the general design code indicate the average stress and strain relationships of concrete in structures with a limited applicable range. Therefore, the stress-and-strain relationship used for the design of concrete structures gives the average stress-and-strain relationship.

The average stress and strain here is generally averaged local stresses and strains in the reference volume element (Representative Volume Element) of the standard size (diameter φ = 100 mm, height H = 200 mm) of the specimen used for compressive strength and tensile strength tests. Similarly, the constitutive laws of concrete under multi-axis stress are used to average the local stresses of different area of concrete surrounded by reinforcing bars in the structures.

Next, about the stress-and-strain relationships of concrete described in the design code, the stress-and-strain relationships on the compression side of the current design code are shown as curves in Figure 2.1 with respect to one axis. Furthermore, this design code describes the tensile softening characteristics related to the stress-and-strain relationships in Figure 2.2.

The horizontal axis and vertical axis of the figure are displayed in a dimensionless value.

This property can evaluate the decrease in tensile stress in the fracture progression region, which is halfway between the elastic region where

Theory of elasto-plastic damage mechanics 9

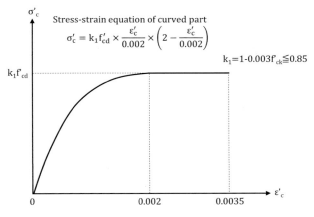

Figure 2.1 Stress-and-strain relationship of concrete (Modified from Japan Society of Civil Engineers 2002, with permission.)

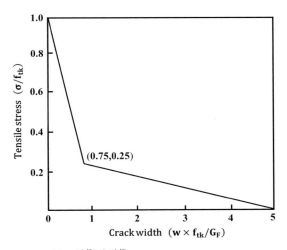

$G_F = 10(d_{max})^{1/3}(f'_{ck})^{1/3}$

G_F: Fracture energy (N/mm)

d_{max}: Maximum size of coarse aggregate (mm)

f_{tk}: Tensile strength of concrete (N/mm^2)

σ: Tensile stress (N/mm^2)

w: Crack width (mm)

Figure 2.2 Characteristics of tensile softening (Modified from Japan Society of Civil Engineers 2002, with permission.)

cracks have not occurred and the cracked part that is fully opened, and by considering this in the analysis, the fracture phenomenon of the concrete associated with crack progress can be evaluated. In addition, it is also considered that the concrete subject to tension can be modelled as a fully brittle material, and the tensile properties of concrete can be incorporated into the structural design.

As described above, the constitutive laws of concrete tend to be maintained so that they can be used in structural analysis, but when examining the deformation leading up to the ultimate state in detail, it is necessary to assume the stress-and-strain relationship in accordance with the softening region rather than the relationship shown in Figure 2.1.

In addition, in order to evaluate the strength and toughness of the member when inspecting seismic performance, it is necessary to introduce the constitutive laws in a biaxial or triaxial stress state. However, it is said that it is affected by many factors such as the quantity of shear reinforcing bars to be constrained, the same deformed bar spacing, axial load, strain rate, etc., and there are still many problems that have not yet been solved for integration into the design model.

In this chapter, since the target structure is also a relatively simple structure, each strength result obtained as a reference volume element is utilized as basic data of the constitutive laws, and the effect of the multi-axis stress state considering the stress state as a member is not considered.

2.3 CONCEPT OF CONTINUUM DAMAGE MECHANICS

Microscopic defects and voids distributed in many of the materials combine with the occurrence and progress, causing macroscopic crack destruction. Continuum Damage Mechanics is a theoretical system that expresses microscopic structural changes by describing such microscopic defects, material degradation due to void growth, such as a decrease in stiffness, strength, toughness, or a decrease in remaining life, in continuous mechanical variables. Macroscopic cracks are often formed by the coalescence of voids. For this reason, damage mechanics have the advantage of consistently modelling the process from the occurrence of voids to the final failure due to the progress of cracks.

In particular, the evaluation of the remaining life and fatigue life of the structural member is considered to be more important in the future for the maintenance technology of social capital, and the evaluation of only the conventional Miner's law (linear cumulative damage law) is insufficient. Damage mechanics that can directly express the development and accumulation of damage by the damage variables, which are the state variables, are considered to be able to take advantage of their features in the evaluation of structural life in particular.

Theory of elasto-plastic damage mechanics 11

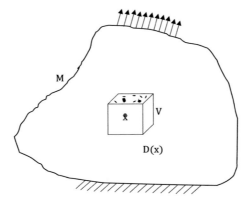

Figure 2.3 Representative volumetric element and damage variables

According to the observation of the electron probe microanalyzer of concrete, concrete generally secures a fine air volume of about 4% of the volume ratio in the concrete mixture. In this observation, it can be confirmed that there are voids around the aggregate and in the cement paste. It can be determined that damage mechanics are relatively easy to apply to concrete with such features.

As shown in Figure 2.3, it is assumed that defects and voids are distributed in the cross-section containing x in the reference volume element V that is sufficiently smaller than M inside object M. In damage mechanics, the damage condition of the element is expressed by appropriate state variables (damage variables). At this time, the volume element V contains enough voids to be uniform, and the stresses and strains of the voids distributed in V, and other state variables can be considered uniform enough.

When the above-mentioned assumption is approved, the problem of damage and destruction by the development of the distributed voids can be developed, in principle, within the frame of continuum mechanics by the following procedure:

(1) Describe the mechanical effect of the voids with the appropriate damage variables D(x)
(2) Formulate the development equations governing the evolution of these damage variables
(3) Create constitutive equations describing the mechanical behaviour of a material damaged by distributed voids
(4) Solve the initial and boundary value problem by these equations

Among the above items, it is necessary to assume what kind of the damage variables are expressed considering mathematical characteristics stage (1). When the anisotropic properties of the materials are remarkable, such as laminated materials, it is necessary to define the damage variables D(x) as a damage tensor. In addition, since metal materials are generally considered as materials that exhibit isotropic mechanical properties, damage variables are treated as scalar quantities. Next, item (2) adopts a format that depends on strain energy release rates in a conjugate relationship with the damage variables in this chapter.

Item (3) will be described in detail in the next section, but the dissipative potential of the system is defined as the sum of the plastic and damage potentials, and finally a constitutive equation is formulated as a tangential elasto-plastic damage tensor. For (4), the finite element method is mainly used considering the characteristics of continuum mechanics.

In addition, the quantification of the damage condition is modelled by reducing the effective cross-sectional area as follows.

As shown in Figure 2.4, it is assumed that the bar member is damaged and deformed by the action of stress σ, and the cross-sectional area changes to A.

A is a total cross-sectional area, which corresponds to the sum of the area excluding the area of the damaged portion such as voids after the change and the area of the damaged portion. Here, the area excluding the damaged portion, such as the void of the former, is referred to as the effective area $\bar{A}$. At this time, the damage variables D can be defined as the decrease rate of the area that effectively resists the external force as the area of the damaged portion increases:

$$D = \frac{A - \bar{A}}{A} = 1 - \frac{\bar{A}}{A} \tag{2.1}$$

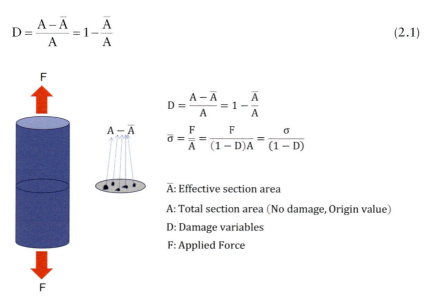

Figure 2.4 Relationship between deformation and damage (Modified from *A Course on Damage Mechanics*, Lemaitre 1990.)

Further, the effective area is expressed from this equation as follows:

$$\bar{A} = (1-D)A \tag{2.2}$$

In addition, the reduction of the effective area expands the effect of stresses due to external forces F. When this effect is defined as effective stresses $\bar{\sigma}$, it is expressed as follows:

$$\bar{\sigma} = \frac{F}{\bar{A}} = \frac{F}{(1-D)A} = \frac{\sigma}{1-D} \tag{2.3}$$

2.4 ELASTO-PLASTIC DAMAGE CONSTITUTIVE LAWS

In applying the concept of effective stresses in the previous section to continuum damage mechanics, it is necessary to assume the relationship between the effective stresses in order to formulate the elasto-plastic damage equations, such as the relationship between the stresses used for the induction of general elastic configuration equations and the plastic potential, the yield function, etc. As this assumption, the following two are proposed:

(1) Strain Equivalence Principle
(2) Energy Equivalence Principle

(1) is a non-damaged material having an effective area $\bar{A}$ excluding a decrease in area due to voids that actually resists external force in the damaged material. Therefore, it is assumed that the deformation of the damaged material is equal to the deformation when the effective stresses $\bar{\sigma}$ act on the non-damaged material of the effective area $\bar{A}$.

On the other hand (2), it is assumed that the thermodynamic potential $w(\sigma, D)$ of Gibbs in the damaged material is equal to replacing the stresses σ in the thermodynamic potential $w_0(\sigma)$ in the non-damaged material with the effective stresses $\bar{\sigma}$. Since the relatively simple strain equivalence principle is frequently used for the development of constitutive laws, this strain equivalence principle is used here. It is described in detail below.

The constitutive equations for the damaged material are based on the stresses σ in the formula of the non-damaged material and derived by replacing it with an effective stress $\bar{\sigma}$. Although the conceptual diagram is shown in Figure 2.5, this strain equivalence principle is very clear and is widely used to formulate the equations of damaged materials so far.

Next, elasto-plastic damage constitutive equations are described.

14 Elasto-plastic damage behaviour of concrete elements

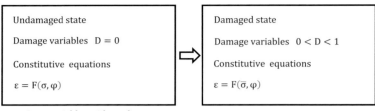

Figure 2.5 Concept of strain equivalence principle (Modified from *A Course on Damage Mechanics*, Lemaitre 1990.)

The dissipative potential of the system F is expressed as the following equation as a sum of plastic potential and damage potential.

$$F = F_p(\bar{\sigma}, \gamma, D) + F_D\{Y;(p,D)\}$$

$$= \overline{\sigma_{eq}} - \gamma - \sigma_y + \frac{S_1}{(S_2+1)(1-D)}\left(\frac{Y}{S_1}\right)^{S_2+1} \quad (2.4)$$

Here, F_p is the potential for the growth of plastic strain and is a function of an effective equivalent stresses $\overline{\sigma_{eq}}$, plastic-hardened parameter and a scalar-damage variables D. In this chapter, the damage variables are assumed to be scalar values, i.e., each material as an isotropic material. F_D is also a damage potential for the evolution of damage and is a function of strain energy release rates Y, equivalent plastic strains p and damage variables D. S_1 and S_2 are material constants.

In the formulation of the constitutive equation and the yield function are assumed as follows:

$$f = \overline{\sigma_{eq}} - \gamma - \sigma_y = 0 \quad (2.5)$$

$$\overline{\sigma_{eq}} = \sigma_{eq}/(1-D) \quad (2.6)$$

$$\sigma_{eq} = \alpha I_1 + \sqrt{J_2'} \quad (2.7)$$

Where the following notations are used:
- $\overline{\sigma_{eq}}$: Drucker-Prager's effective equivalent stresses
- σ_y: The yield stress
- α: Material parameter
- I_1: The first invariant of stress
- J_2': The second invariant of stress deviator

Theory of elasto-plastic damage mechanics 15

In the plastic state, the yield function of Equation (2.5) is the plastic potential, and the following equation is established on the yield phase including damage:

$$dF_p = \left(\frac{\partial F_p}{\partial \bar{\sigma}}\right)^T d\bar{\sigma} + \left(\frac{\partial F_p}{\partial \gamma}\right) d\gamma + \left(\frac{\partial F_p}{\partial D}\right) dD = 0 \qquad (2.8)$$

In addition, plastic strain increments d_p and equivalent plastic strain increments dp are represented by the following equations:

$$d\varepsilon_p = d\lambda \frac{\partial F}{\partial \bar{\sigma}} = d\lambda \frac{\partial F_p}{\partial \bar{\sigma}} \qquad (2.9)$$

$$dp = -d\lambda \frac{\partial F_p}{\partial \gamma} \qquad (2.10)$$

Here, $d\lambda$ is the proportional coefficient.

Since the total strain increment in the plastic state is the sum of the elastic strain increments and the plastic strain increments, the effective stress increments are represented by the following equations.

$$d\bar{\sigma} = Cd\varepsilon_e = C(d\varepsilon - d\varepsilon_p) = Cd\varepsilon_p - Cd\lambda \frac{\partial F_p}{\partial \bar{\sigma}} \qquad (2.11)$$

C is a general stress-and-strain matrix of six rows and six columns, and $d\varepsilon_e$, $d\varepsilon_p$ are elastic and plastic strain increments, respectively. Further, the plastic-hardened parameters and their increments are assumed as follows:

$$\gamma = Kp^n \qquad (2.12)$$

$$d\gamma = nKp^{n-1}dp = Hdp \qquad (2.13)$$

K and n are material constants.

Further, the increments of the damage variable are obtained by the following equation:

$$dD = d\lambda \frac{\partial F_D}{\partial Y} \qquad (2.14)$$

When Equations (2.14), (2.11), and (2.13) are arranged by assigning them to Equation (2.9), the proportional coefficient $d\lambda$ is calculated as follows:

$$d\lambda = \frac{\left(\frac{\partial F_p}{\partial \bar{\sigma}}\right)^T C}{H + \left(\frac{\partial F_p}{\partial \bar{\sigma}}\right)^T C \frac{\partial F_p}{\partial \bar{\sigma}} - \frac{\bar{\sigma}_{eq}}{1-D} \frac{\partial F_D}{\partial Y}} d\varepsilon \qquad (2.15)$$

16 Elasto-plastic damage behaviour of concrete elements

When Equation (2.15) is assigned to Equation (2.11), the relationship between the effective stress increments and the strain increments are obtained as follows:

$$d\bar{\sigma} = C\left[1 - \frac{\frac{\partial F_p}{\partial \bar{\sigma}}\left(\frac{\partial F_p}{\partial \bar{\sigma}}\right)^T C}{H + \left(\frac{\partial F_p}{\partial \bar{\sigma}}\right)^T C \frac{\partial F_p}{\partial \bar{\sigma}} - \frac{\sigma_{eq}}{1-D}\frac{\partial F_D}{\partial Y}}\right]d\varepsilon \qquad (2.16)$$

The relationship between stress increments and strain increments are expressed by the following equations:

$$d\sigma = (1-D)d\bar{\sigma} - dD\bar{\sigma} = D_{epd}d\varepsilon \qquad (2.17)$$

$$D_{epd} = \left[(1-D)C - \left\{(1-D)C\frac{\partial F_p}{\partial \bar{\sigma}} + \bar{\sigma}\frac{\partial F_D}{\partial Y}\right\}\frac{\left(\frac{\partial F_p}{\partial \bar{\sigma}}\right)^T C}{H + \left(\frac{\partial F_p}{\partial \bar{\sigma}}\right)^T C \frac{\partial F_p}{\partial \bar{\sigma}} - \frac{\sigma_{eq}}{1-D}\frac{\partial F_D}{\partial Y}}\right]$$

(2.18)

Here, D_{epd} is a tangential stress-and-strain matrix that considers elasto-plastic damage that relates stress increments and strain increments. Since this matrix is an asymmetric tensor with six rows and six columns, the attention should be paid to the solver selection for the analyses.

In the above process, the elasto-plastic damage constitutive equations are formulated.

Next, the damage evolution equation basically uses the following equation. It is assumed that the damage progresses with the increase of equivalent plastic strains.

$$dD = \left(\frac{Y}{S_1}\right)^{S_2} dp \qquad (2.19)$$

In addition, the strain energy release rates Y are represented by Young's modulus E and the equivalent stresses σ_{eq} as follows:

$$Y = \frac{\sigma_{eq}^2}{2E(1-D)^2} \qquad (2.20)$$

Chapter 3

Strength characteristics of concrete by elasto-plastic damage mechanics

3.1 BACKGROUND

In the design of concrete structures, it is necessary to set compressive strength, tensile strength, or bending strength in all cases, although structural analysis by experimental empirical formula or finite element method, etc., is used. In general, the compressive strength of concrete is designed based on the design reference strength and defined as a normal concrete in the range of 20 to 50 (N/mm^2), a range exceeding 50 (N/mm^2) as a high-strength concrete. Recently, high-strength concrete with design reference strength exceeding 100 (N/mm^2) has also been used to ensure the seismic performance required for piers and skyscrapers over 50 metres high.

The compression, tensile, and bending strength of ordinary concrete have accumulated a large number of experimental templates so far, so tensile and bending strength are often estimated based on compressive strength. However, these are estimates of the strength used in the design, and do not present specific constitutive laws used for structural analyses. In particular, the tensile and bending strength are as small as about 1/6 to 1/13 of the compressive strength and are important data for evaluating the strength of the concrete members.

This chapter first shows the material strength test results of single-axis compression and tensile performed to formulate the elasto-plastic damage constitutive equations used in finite element analysis. The compressive strength test was based on the "JIS A 1108 concrete compressive strength test method" and bending strength test was based on the "JIS A 1106 concrete bending strength test method".

The tensile strength tests were not conducted by a splitting tensile test method using the "JIS A 113 concrete tensile strength test method", but by direct tensile test. Since the splitting tensile tests tend to evaluate the tensile strength by about 10%, the direct tensile tests were performed to ensure accuracy. In addition, since the bending strength test has relatively few experimental results and are effective data for evaluating the bending fatigue characteristics of the member, the bending fatigue strength tests were also conducted.

DOI: 10.1201/9781003284246-3

18 Elasto-plastic damage behaviour of concrete elements

By the above various experiments, stress-strain curves and load displacement curves, which are the strength characteristics of concrete, were obtained. Referring to these results, the material constants in the elasto-plastic damage constitutive laws were determined.

3.2 STATIC COMPRESSIVE STRENGTH

3.2.1 Experimental results

Table 3.1 shows the mixture proportion of concrete.

This proportion was set to be equivalent to the compressive strength of the specimen sampled directly from the actual bridge to be studied.

The loading equipment was a type of computer-controlled and hydraulic servo motor.

The specimens for single-axis compression and tensile strength tests were performed using a cylindrical specimen having a diameter of φ100 (mm) and a height of 200 (mm). In addition, a strain gauge for concrete on the side of the specimen was pasted to measure the strains.

Figure 3.1 shows the stress curve and the strain curve by the single-axis compressive strength tests. In addition, the material age is a general 28 days. Note that Ex.1 and Ex.2 in the legend show experimental data.

Since the stress and strain are positive on the tensile side, the compression side is negative, but here the stress-and-strain curves are graphed as positive

Table 3.1 The mix proportion of concrete

G_{max} (mm)	W/C (%)	Water	Cement	Fine aggregate	Coarse aggregate	Admixture
25	62	152	243	843	1013	2.43

Unit weight (kg/m³) spans Water through Admixture.

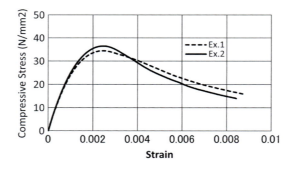

Figure 3.1 Stress-and-strain relationships (compression)

Table 3.2 Material parameters (compressive side)

Young's Modulus [1]	29,700 (N/mm²)
Poisson's Ratio	0.2
Compressive strength (maximum)	34.5 (N/mm²)
Maximum strain	2,400 (μ)

1) This value is the slope of the straight line connected to the stress-and-strain point of 1/3 point of compressive strength and the origin.

to confirm the strain-softening region. Table 3.2 shows the material parameters obtained by compression strength tests. Each parameter is the average value of the results. The stress-and-strain relationships in compression are good results.

Figure 3.2 shows the situation before the compressive strength tests and the situation after tests, respectively. The figure schematically shows the concept of the fracture process. It is thought that the cracks in the test piece load direction are generated by the tensile stress in the circumferential direction.

The compressive fracture can be expressed in a model in which cracks generated at the aggregate interface grow in mortar and eventually connect, and the aggregate of columnar concrete masses surrounded by cracks exhibit buckling modes and progress throughout.

The compressive fracture process of concrete can be thought of as a collection of microscopic tensile fractures, and tensile cracks are thought to open in the direction with the least binding force. In addition, this fracture process is considered to be the mixed mode of modes I and II of fracture mechanics.

3.2.2 Stress-strain curve for compressive side

The stress-and-strain relationships on the compression side of the concrete are identified from the elasto-plastic damage constitutive laws and single-axis compressive strength tests described in Chapter 2. Next, the process of identifying the stress-and-strain relationships obtained by the compression strength tests and the laws are described.

Elasto-plastic damage constitutive equations in the incremental form:

$$d\sigma = D_{epd} d\varepsilon \quad (3.1)$$

$$\begin{Bmatrix} d\sigma_1 \\ d\sigma_2 \\ d\sigma_3 \end{Bmatrix} = \begin{bmatrix} D_{epd11} & D_{epd12} & D_{epd13} \\ D_{epd21} & D_{epd22} & D_{epd23} \\ D_{epd31} & D_{epd32} & D_{epd33} \end{bmatrix} \begin{Bmatrix} d\varepsilon_1 \\ d\varepsilon_2 \\ d\varepsilon_3 \end{Bmatrix} \quad (3.2)$$

20 Elasto-plastic damage behaviour of concrete elements

(a) Before test

(b) After test

(c) Cracking process

Figure 3.2 Compressive strength test and final state

Here, $d\sigma_1$, $d\sigma_2$, $d\sigma_3$ are the principal stress increments, and the corresponding strain increments $d\varepsilon_1$, $d\varepsilon_2$, $d\varepsilon_3$ are the principal strains.

Since the behaviour in the single-axis direction of the material strength test is dominant, the principal stresses in the other biaxial direction are zero. In addition, the elasto-plastic damage tensor is asymmetric.

Here, if $d\sigma_1$ is the principal stress direction of the single-axis compression, the stress increments ($d\sigma_2$, $d\sigma_3$) of the other biaxial component are zero, and the strain increments of the biaxial component ($d\varepsilon_2$, $d\varepsilon_3$) can be defined in linear functions by $d\varepsilon_1$.

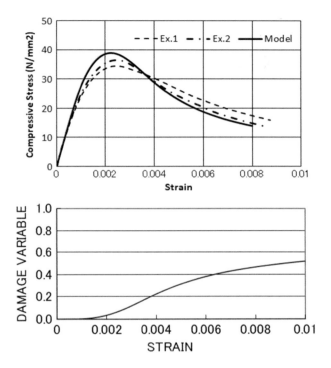

Figure 3.3 Identification of stress-and-strain relationships and damage variables (compression)

The material parameters are set by identifying the stress-and-strain relationships obtained by the stress and strain calculated by organizing the above relationships and the stress and strain by the single-axis compressive strength test.

The stress-and-strain curves identified in Figure 3.3 are shown with the experimental results. The distribution of damage variables is also posted in the figure. Ex.1 and Ex.2 in the figure show experimental results, and Model shows the results identified. The identified results are consistent with the experimental results up to the strain-softening region after compressive strength. As for the damage variable, the compressive stress has increased exponentially from around 20 (N/mm²), but after the softening region, it tends to gradually increase to about 0.6. Table 3.3 shows the material parameters of concrete determined when identifying the stress-and-strain relationships obtained from the compressive strength tests. These parameters are also applied to the stress-and-strain relationship obtained from the tensile strength tests described later.

The stress-and-strain relationships, including repeated unloading, are shown in Figure 3.4. The vertical axis of this figure is standardized by compressive strength f_c', and the horizontal axis is by strain ε_m at the time

Table 3.3 Material parameters for constitutive laws (concrete)

Young's Modulus (E)	29,700 (N/mm²)
Poisson's Ratio (ν)	0.2
Coefficient of equivalent stress (α)	0.72
Yield stress (σ_y)	0.75 (N/mm²)
Coefficient of plastic hardening (K)	40.0 (N/mm²)
Coefficient of plastic hardening (n)	0.215
Coefficient of damage evolution (S_1)	0.215×10^{-3} (N/mm²)
Coefficient of damage evolution (S_2)	1.55
Damage threshold (ε_{pd})	0.00
Critical damage variable (D_{cr})[1]	2.75×10^{-5}

1) This value is used as a limit value at the time of breaking that causes cracks in the tensile region.

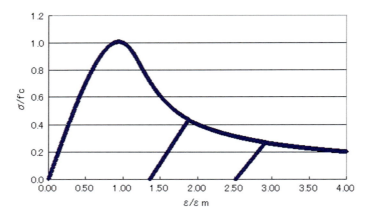

Figure 3.4 Reference stress-and-strain relationships by identification models

of compressive strength. The straight portion after unloading becomes E (1−D), which represents a decrease in the elastic modulus associated with the damage variables. Further, the strain at the time of complete unloading represents plastic strain, and it can be estimated that damage and plasticity increase with the load from the second unloading behaviour.

3.3 STATIC TENSILE STRENGTH

3.3.1 Experimental results

Next, for tensile strength tests, the splitting tensile strength test method generally shown in Figure 3.5 is defined in the Japanese Industrial Standard (JIS) using the same cylindrical specimens in the compressive strength tests.

Characteristics of concrete by elasto-plastic damage mechanics 23

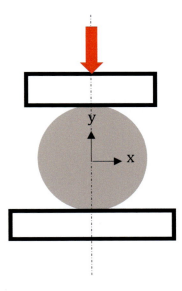

Figure 3.5 Splitting tensile strength test method (JIS)

This strength is due to the tensile stresses generated in the x direction of the figure, but at the same time the compressive stresses occur in the y direction and the stress state in the cross-section is complicated. In addition, this method is relatively simple, and stable tensile strength can be obtained indirectly, but since the stress-and-strain relationship cannot be measured accurately, these were measured by direct tensile strength tests in Figure 3.6. In these tests, it may be affected by the bending moment at the time of the tests, but they were evaluated by direct tensile tests as well as the tensile strength tests of the steel material. The mix proportion of concrete for tensile strength tests is the same as for compression strength tests.

The figure shows the situation before the tensile strength tests, and after the tests. In the tensile strength tests, up to the maximum stress could be measured, but the subsequent softening area was impossible to measure due to rapid brittle fracture. The average tensile strength was 2.11 (N/mm²) and the average maximum strain was 109 (μ) in Figure 3.7.

In the figure, Ex.1 and Ex.2 show experimental data, respectively. In these tests, it was found that cracks occur perpendicular to the loading direction, which rapidly cross the section and exhibit brittle behaviour.

3.3.2 Stress-strain curve for tensile side

Figure 3.8 shows the tensile stress-and-strain relationship. The stress-and-strain relationships before breaking are presented assuming that it breaks at the time of stress corresponding to the average tensile strength by

24 Elasto-plastic damage behaviour of concrete elements

Figure 3.6 Direct tensile strength test and final state

experiments. The model in the figure shows the stress-and-strain relationships using the constitutive rules, and the parameters of the rules use the same values as the compression side (refer to Table 3.3).

The figure shows the damage variable and strain relationships. From this figure, the strain corresponding to the tensile strength is the breaking strain, and the damage variable corresponding to this breaking strain is set as the critical damage value (Dcr). In the numerical analysis described later, crack progress is expressed by reducing the stress to zero when the tensile stress or strain reaches the damage value (Dcr).

In addition, it is judged that the stress-and-strain relationships are in agreement with the experimental results as well as the compressive side.

This constitutive equation applied equivalent stress of Drucker–Prager's law, but it is able to identify the stress-and-strain relationships of concrete whose compressive strength and tensile strength are more than ten times different.

Characteristics of concrete by elasto-plastic damage mechanics 25

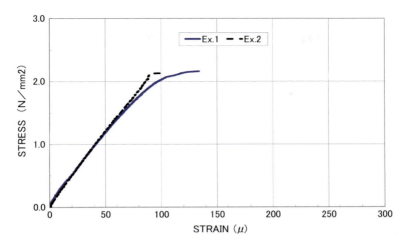

Figure 3.7 Stress-and-strain relationships (tension)

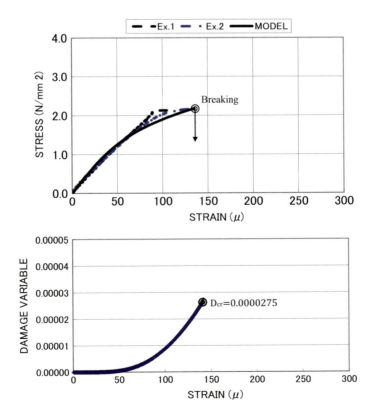

Figure 3.8 Identification of stress-and-strain relationships and damage variables (tension)

26 Elasto-plastic damage behaviour of concrete elements

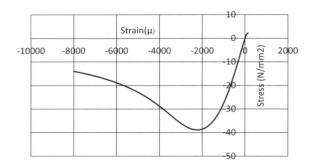

Figure 3.9 Stress-and-strain relationships between compression and tensile

Figure 3.9 shows the stress-and-strain relationships between compression and tensile which are identified. Although the rapid strain-softening phenomenon after tensile strength cannot be expressed, the stress-and-strain relationships of concrete with compressive strength and tensile strength more than ten times different can be evaluated.

3.4 STATIC BENDING STRENGTH

3.4.1 Experimental results

In the bending strength test of the concrete, a rectangular shape of 150 (mm) × 150 (mm) and a prismatic column with a length of 600 (mm) is three equal point loaded. Figure 3.10 shows the loading and support parts of the bending loading steel jig. The loading part is for two-point loading, and two steel plates with a thickness of 30 (mm) are welded, respectively, and the distance between the two-point loading is 150 (mm). The steel plate is bolted and connected to the jack with a rotating hinge in a plane. Ring steel (200 (mm) deep) of φ 30 (mm) is installed at the tip of the loading point, and the contact point with the specimen is rotatable and friction is reduced.

Further, steel plates having a thickness of 30 (mm) and 25 (mm) are welded to an L shape, and further bolted to steel plates with a thickness of 25 (mm) and a length of 650 (mm) at a centre distance of 450 (mm). The tip of the support portion is equipped with φ30 (mm) ring steel as well as the tip of the loading part, and rotation is free.

The mix proportion of concrete for bending strength tests is the same as for compression strength tests (refer to Table 3.1).

Figure 3.11 shows the load and central displacement curves. The load and centre point displacement curve showed almost elastic behaviour up to the maximum load, and thereafter showed a brittle fracture that rapidly reduced bearing capacity with the occurrence and progress of cracks.

Figure 3.12 shows the situation after bending strength tests. It was found that the cracks in the bending strength tests were scattered near the central

Characteristics of concrete by elasto-plastic damage mechanics   27

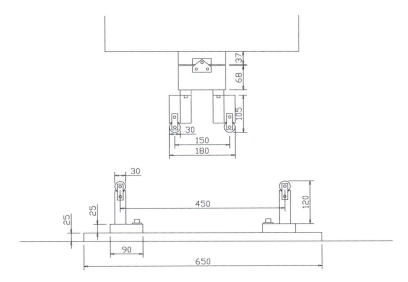

Figure 3.10 Steel jig for bending strength test (unit: mm)

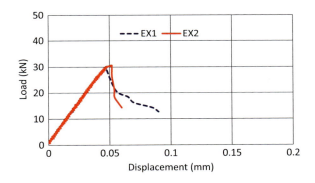

Figure 3.11 Load and central displacement curves

portion in the bending section and around both loading points depending on the material properties of the concrete. Since the cracks generated in each test specimen had progressed in the same way to the thickness direction of the specimen, it is considered that there was little effect by torsion.

3.4.2 Stress release of crack elements

When analyzing from the formulation of the finite element method to the analysis of the elasto-plastic damage constitutive equations concretely into a finite element programme, the macro-crack evaluation method is important to represent an algorithm that causes meso-cracks beyond the critical

28 Elasto-plastic damage behaviour of concrete elements

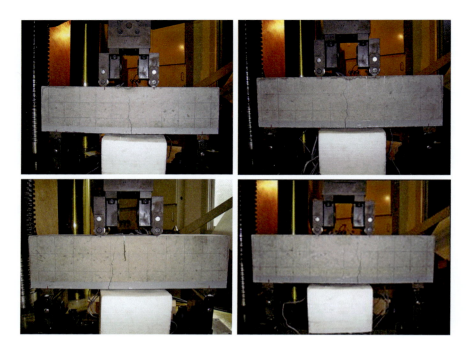

Figure 3.12 Crack state of bending test pieces

damage value and develops into macro-cracks and progresses to material failure or structural collapse.

In general, the strain-softening region can be analyzed accurately by arc length method, perturbation method, and modified Newton–Raphson method, etc., but the stress of a brittle material that rapidly decreases in stress after a strength like concrete may be difficult to analyze accurately.

Therefore, here, the processing method in finite element analysis when macro-cracks occur (these are called crack elements) is summarized.

In damage mechanics, when the damage variables D reach the critical damage value Dcr, the element is considered a crack element and can be approximated to represent macro-cracks.

It is necessary to set the following two conditions for the mechanical characteristics of the crack elements.

(1) There is no resistance performance to external force and external force increments in analysis after crack generation. That is, the element rigidity is zero. It cannot be completely zero in the calculation, so it is assumed that it is close to zero which does not become an engineering problem
(2) Releases internal energy stored prior to cracking. That is, the stresses of the crack elements are released

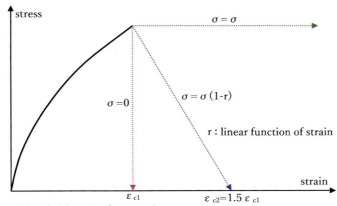

ε_{c1}: Threshold strain of stress release
ε_{c2}: Fully release strain

Figure 3.13 Stress release of crack elements

(1) is as close as possible to 1.0 when the damage variables D exceed Dcr.

By considering the value (D = 0.99 in the calculation), Equation (3.2) approximates zero,

The tangential stiffness tensor K_t of Equation (2.28) can also approximate zero.

As for (2), in brittle materials such as concrete, the stress in the tensile state is rapidly released. Figure 3.13 shows an example of releasing stress.

3.4.3 Analysis of static bending strength

In this section, I show the simulation results of the monotonic loading and bending strength test of the concrete. The analysis is the plane strain element and is presented in comparison with the load and central point displacement curves obtained in the tests.

Figure 3.14 shows the mesh division used in finite element analysis. The supports are simple support condition. In addition, the elasto-plastic damage constitutive laws are the same as the static tensile tests. Although it is possible to identify the elasto-plastic damage constitutive laws for bending strength, when the same constitution rules are applied to numerical analysis of concrete structures, it is not practical to select the rules for each deformation characteristic such as bending deformation and shear deformation, and the scope of application is limited. In consideration of the above, the constitutive rules for tensile strength are evaluated by applying them to bending strength characteristics.

Figure 3.15 shows the load-central point displacement curves with the experimental results. The maximum load of the experimental results is about 1.3 times the analysis value. In general, the strength obtained by

30 Elasto-plastic damage behaviour of concrete elements

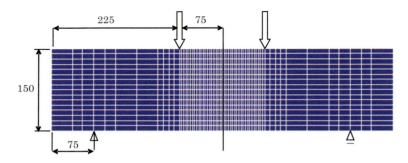

Figure 3.14 Mesh division used in finite element analysis (unit: mm)

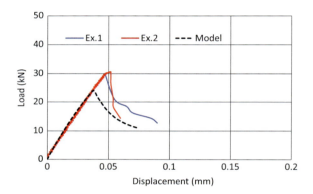

Figure 3.15 Identification of load and central displacement curves comparing with the experimental results

bending strength tests based on JIS is said to be 1.2 to 1.5 times the tensile strength, and this analysis result using the constitutive law of tensile strength seems to be appropriate. Furthermore, it should be noted that the tangential gradient corresponding to the stiffness of the curve before the maximum load of the figure is well in agreement with the experiments and the analyses. The analysis results assume the rapid brittle fracture that releases stresses when the damage variables reach the critical value is consistent with the experiment after the maximum load. Legends Ex.1 and Ex.2 in the figure show the experimental results and the Model shows the analysis results.

Figure 3.16 shows the distribution of damage variables. This analysis introduces the concept of crack elements that release stresses to express the progress of cracks and brittle fractures when the damage variables reach a critical point. The cracks progress almost linearly toward the top surface from the lower surface, which is the tensile region in the central part of the

Characteristics of concrete by elasto-plastic damage mechanics 31

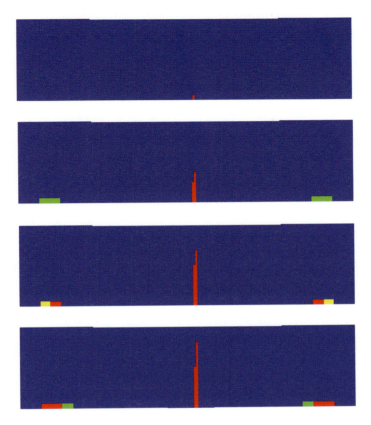

Figure 3.16 Distribution of damage variables compatible with load increments

test piece. The distribution of the critical damage variables is well matched with the crack distribution of the experiments.

Figure 3.17 also shows the distribution of equivalent stresses σ_{eq}(N/mm^2). These stresses are distributed almost evenly in the tensile part on the underside of the test piece before the crack generation. However, when the damage variables reach the critical value, it is possible to visualize that the stresses are released with the progress of the cracks.

It can be presumed that the stress at the crack tip is larger than that in other regions, and that the characteristics of the cracks can be expressed.

3.5 BENDING FATIGUE STRENGTH

Concrete is generally evaluated for each strength by static loading tests. However, it is insufficient to perform structural design only by these values. In particular, dynamic strength considering inertial and viscous terms

32 Elasto-plastic damage behaviour of concrete elements

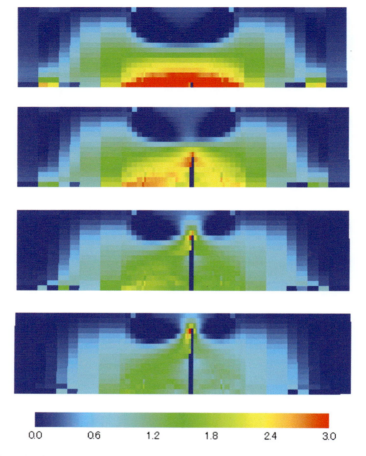

Figure 3.17 Distribution of equivalent stresses compatible with load increments (unit: N/mm²)

and fatigue strength are important characteristics in order to evaluate load resistance and life.

In this section, we explain the bending fatigue strength characteristics that are rarely performed by experiments and numerical analyses.

3.5.1 Experimental results

The bending fatigue strength tests were carried out using the same prism test pieces. The loading was conducted with the actuator manufactured by MTS System Corporation by the load control method.

The loading method was performed by a three-iso-point bending load as well as the bending strength tests, and the complete swing of all cases and

Table 3.4 Maximum repeated load and fatigue failure frequency

Maximum load (kN)	Failure frequency
36.0	146
32.3	1,383
30.0	7,132
28.0	90,187
24.6	1,037,477

2 (Hz) cycle were made. In addition, in so that data from low cycle of about 100 times to high cycle failure of about 1,000,000 times was obtained, the ratio to the maximum load obtained by the static bending strength was changed and the load was set repeatedly. Table 3.4 shows the maximum repetition load and fatigue failure frequency.

It can be determined that the relationship between the maximum repetition load and the fatigue number (logarithmic display) in Figure 3.18 obtained relatively good results. However, it can be understood from Figure 3.19 that the position of the main crack generated in the final fracture stage depends on the influence due to the variation of the material as well as the bending strength tests. In all fatigue failure experiments, the specimen has reached a bending fracture at the same time as cracking occurs when the number of fatigue fractures is reached.

The load and central point displacement curves are shown in Figure 3.20. These are results in which the number of fatigue fracture corresponds to 90,187 times. As the number of fatigues increases, the decrease in the gradient of the curve is common to all results. This can be interpreted as the damage progresses with increasing fatigue, resulting in an effective cross-sectional area resisting external forces or a decrease in bending stiffness, which is the gradient of the curve.

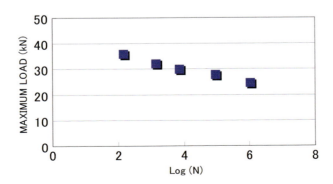

Figure 3.18 Relationship between maximum repetition load and fatigue failure frequency

34 Elasto-plastic damage behaviour of concrete elements

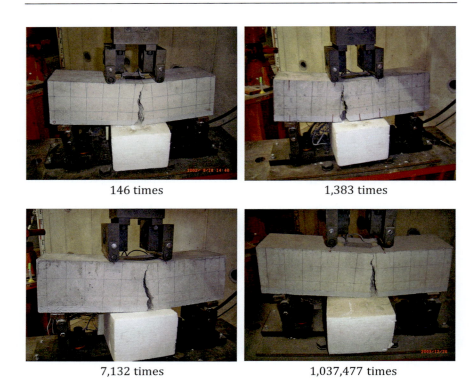

Figure 3.19 Number of fatigue failures and fracture situation

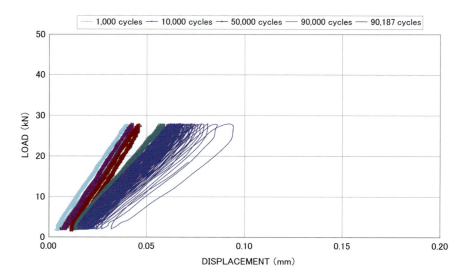

Figure 3.20 The load and central point displacement curves for repetition

This decrease in bending stiffness is the phenomenon in which the tangent gradient in the stress-and-strain relationships is caused by an increase in the damage variables according to the number of fatigue cycles, consistent with the fatigue failure characteristics defined by continuum damage mechanics.

In particular, it can be understood that the features of the curve close to the number of fatigue failures appear remarkable. The load and central point displacement curves of the number of fatigues shown as a legend of each result are data from about 50 times before the fatigue frequency presented.

In these fatigue tests, the minimum value of the repeated loading load is about 2 (kN). This is a measure to prevent the loading point from shifting with the increase in the number of repetitions when returning to the origin, and in fact it is judged that the minimum value can be the origin.

3.5.2 Bending fatigue strength analyses

In this book, the fatigue damage development equation is evaluated by dividing it into elastic damage and plastic damage. That is, when the effective equivalent stress $\overline{\sigma_{eq}}$ exceeds the fatigue limit stress σ_f even in the elastic state of the damage at the time of fatigue, it is assumed that the fatigue damage progresses in proportion to the equivalent elastic strain.

At elastic state,

$$dD_e = \left(\frac{Y}{S_{ef1}}\right)^{S_{ef2}} de \quad (3.3)$$

dD_e: Increments of elastic damage variable
Y: Strain energy release rates
S_{ef1}, S_{ef2}: Material constants for elastic
de: Equivalent elastic strain increments

At plastic state,

$$dD_p = \left(\frac{Y}{S_{pf1}}\right)^{S_{pf2}} dp \quad (3.4)$$

dD_p: Increments of plastic damage variable
S_{pf1}, S_{pf2}: Material constants for plastic
dp: Equivalent plastic strain increments

Ultimately, the damage variables at the time of fatigue are defined as the arithmetic sum of both. Table 3.5 shows the parameters of fatigue damage. Here, the parameters of fatigue damage are described separately into

Table 3.5 Material constants for bending fatigue

σ_f **(Fatigue stress)**	1.3 (N/mm²)
S_{pf1} **(Damage parameter for plastic)**	0.215×10^{-3} (N/mm²)
S_{pf2} **(Damage parameter for plastic)**	1.55
S_{ef1} **(Damage parameter for elastic)**	0.355×10^{-3} (N/mm²)
S_{ef2} **(Damage parameter for elastic)**	8.75

monotonous and fatigue loads. For the identification of stress-and-strain relationships during fatigue, the data obtained in the bending fatigue strength tests in the previous section are used.

For fatigue-bending failure analysis, the relationships between stress and strain are used. The analyses are performed using Fully Coupled Analysis and Locally Coupled Analysis in this instance.

But for brittle materials such as concrete, cracks often mean structural collapse, so Locally Coupled Analysis, which can reduce the calculation time, is effective. For Fully Coupled Analysis, the crack elements are subject to a condition of stress release when the damage variables reach the critical value D_{cr}.

In the bending fatigue fracture analysis, Fully Coupled Analysis and Locally Coupled Analysis were performed. Fully Coupled Analysis is a method to model the overall structure and calculate the physical quantity of all elements in the full load step until the structure is destroyed, and while the accuracy is good, there are problems that require calculation time and cost. These problems are solved to some extent with the improvement of computing techniques, but practical analysis methods are limited to the application of basic structural models and low cycle fatigue.

On the other hand, Locally Coupled Analysis extracts an element to be paid attention to in all structural members or stress-and-strain relationships at integral points by Fully Coupled Analysis for only a few cycles. Then, this relationship can be input into a one-dimensional fatigue model as defined in Equation (3.1) to calculate cumulative damage variables up to the number of fatigue fractures and a predetermined number of fatigues by inputting this relationship into constitutive laws in this chapter. In addition, this analysis method can save calculation time and cost more than the Fully Coupled one. Furthermore, since the fatigue fracture behaviour of materials such as concrete, ceramics, and rock is brittle, and local cracks generated at the time of fatigue fracture often progress to the fracture and collapse state of the entire structure almost simultaneously, it can be expected as an effective analysis method for evaluating structural life and local cumulative pre-damage.

The results of Locally Coupled Analysis are described below. Figure 3.21 is a diagram representing the mesh division of a Fully Coupled Analysis and the representative element and an integration point of interest in the element.

Characteristics of concrete by elasto-plastic damage mechanics 37

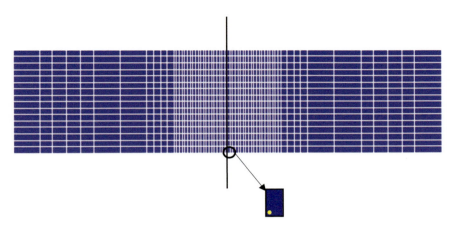

Figure 3.21 Representative integrating point within structural elements to focus on

Here, the central point of the lowest end of specimen is the focus point of interest. The stress amplitude or strain amplitude, which is the response value of the point of interest for several cycles obtained by Fully Coupled Analysis, is extracted. Next, the repetition frequency until the damage variables reach the critical value D_{cr} is calculated using this response value as an input data point of the one-dimensional fatigue model. Then, this is defined as the number of fatigue fractures. The above is a rough procedure for Locally Coupled Analysis.

Figure 3.22 shows an example of an input strain amplitude of Locally Coupled Analysis by approximating the strain amplitude of Fully Coupled Analysis obtained for ten cycles.

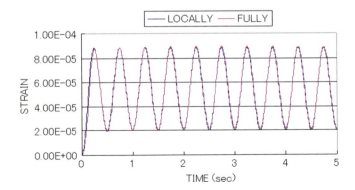

Figure 3.22 Strain amplitude of the focusing integral point obtained from Fully Coupled Analysis

38 Elasto-plastic damage behaviour of concrete elements

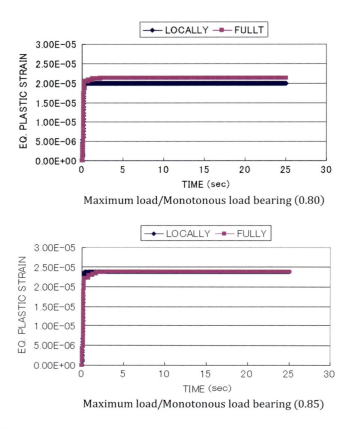

Figure 3.23 Cumulative equivalent plastic strain of the integral point

Figure 3.23 shows examples of the cumulative equivalent plastic strain distribution comparing Fully Coupled Analysis and Locally Coupled Analysis when the maximum load and monotonous load-bearing ratio are 0.85 and 0.80. Furthermore, Figure 3.24 shows examples of the cumulative damage variable distribution corresponding to Figure 3.23.

Since the maximum load and the monotonic load-bearing ratio increase, it means low cycle failure, so when this ratio is large (0.85), it can be seen that the cumulative values of equivalent plastic strain and damage variable by Fully Coupled Analysis and Locally Coupled Analysis are well matched.

Next, Table 3.6 shows the number of load and fatigue fractures that have been made dimensionless by the strength of the monotonous loading and bending failure experimental load considering the results of bending fatigue failure analysis by Fully Coupled Analysis and Locally Coupled Analysis. When the Locally Coupled Analysis and the Fully Coupled Analysis are

Characteristics of concrete by elasto-plastic damage mechanics 39

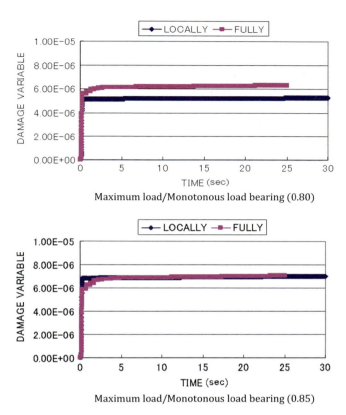

Figure 3.24 Cumulative damage variable of the integral point

compared, the number of fatigue fractures by the Locally Coupled Analysis is evaluated more widely. As a cause, in the Fully Coupled Analysis, plastic deformation progresses during the initial repetition, while in the Locally Coupled Analysis, the plastic deformation has hardly progressed compared to the Fully Coupled Analysis, so it can be determined that the accumulation

Table 3.6 List of times of bending fatigue failure

Fully Coupled Analysis	Locally Coupled Analysis	Experiments
140(0.95)	216(0.95)	146(0.95)
1,962(0.85)	1,436(0.85)	1,383(0.85)
6,762(0.80)	12,696(0.80)	7,132(0.80)
Without implementation	92,752(0.77)	90,187(0.74)
Without implementation	919,772(0.68)	1,037,477(0.65)

Parentheses show the ratio of the maximum load (bearing capacity) in the monotonous bending strength test to each repeatedly maximum load.

40   Elasto-plastic damage behaviour of concrete elements

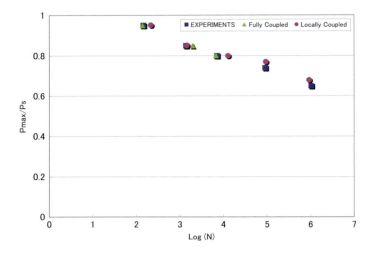

Figure 3.25 Load-fatigue failure frequency relationship

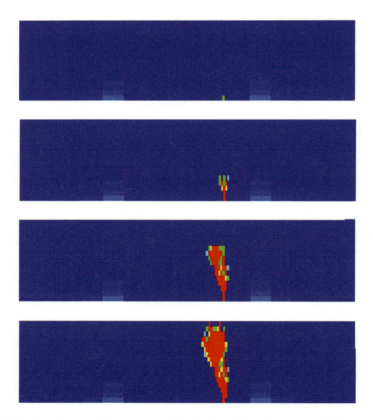

Figure 3.26 Distribution of damage variables (fatigue failure frequency 1,962)

Characteristics of concrete by elasto-plastic damage mechanics 41

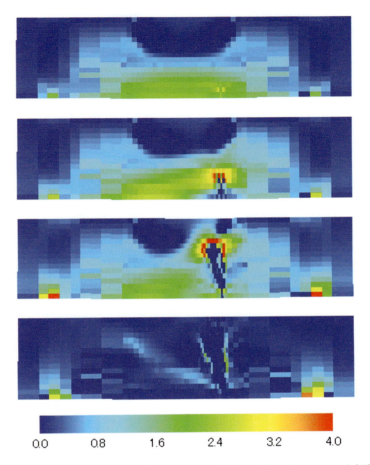

Figure 3.27 Distribution of equivalent stresses (fatigue failure frequency 1,962, unit: N/mm²)

of the damage variables due to this plasticity influenced the difference in the number of fatigue fractures. This can be estimated from the cumulative equivalent plastic strains of Figure 3.23 and the cumulative damage variable distributions of Figure 3.24.

However, from the load and fatigue failure frequency relationship of Figure 3.25, it can be determined that the experimental results from a low cycle of about 100 cycles to a high cycle region of 1,000,000 cycles and both analysis results correspond well.

Figure 3.26 shows the distribution of damage variables obtained by Fully Coupled Analysis at 1,962 fatigue failure times. This distribution results in the macro-cracks generated at the tensile edge of the bending

42 Elasto-plastic damage behaviour of concrete elements

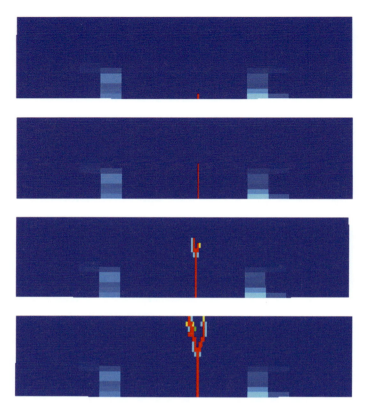

Figure 3.28 Distribution of damage variables (fatigue failure frequency 6,762)

specimen during fatigue failure coinciding with the experimental result of rapidly penetrating the member. Figure 3.27 also shows the corresponding stress distribution at the time of fatigue failure. As with the simulation of monotonic loading strength tests, stress release and stress concentration around the tip associated with the macro-crack progress are observed.

Figure 3.28 shows the damage variable distribution obtained by Fully Coupled Analysis at 6,762 fatigue failure times, and the corresponding stress distributions are shown in Figure 3.29. The features of both are the same as the result of the fatigue failure frequency 1,962. However, it was found that the width and generation position of the macro-crack elements differ depending on the number of fatigue fractures due to the influence of the size of the loads. Although it is not a result seen in all Fully Coupled Analyses, in elements where compressive stresses and tensile stresses are mixed, cracks diverge from the middle as shown in

Characteristics of concrete by elasto-plastic damage mechanics 43

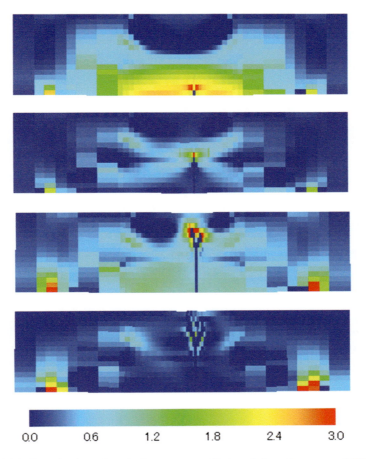

Figure 3.29 Distribution of equivalent stresses (fatigue failure frequency 6,762, unit: N/mm²)

Figure 3.28. For phenomena in which the generation position and progress of such crack elements differ depending on the stress state, it seems to depend mainly on the accuracy of the constitutive laws when these stresses are mixed.

Figure 3.30 describes the experimental and Fully Coupled Analysis results about the load and central point displacement curves. The 500, 1,000, and 1,383 cycles in the figure represent the experimental results, and 1,962 cycles represent the curves of the analysis results, displaying the results from the previous 50 cycles of each number.

Both the experimental and the analytical results tend to increase plastic deformation according to the fatigue frequency. However, the experimental results show that the gradient of the curves decrease according to

44 Elasto-plastic damage behaviour of concrete elements

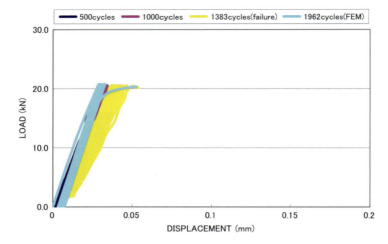

Figure 3.30 Comparison of load-central displacement curve with experiment (fatigue failure frequency 1,383 by the experiments)

the number of fatigues, but the tendency is not remarkable in the analysis. These changes in gradient are thought to mainly show the development of the damage variables, and further improvement of the damage development equation with fatigue seems to be necessary to harmonize with the experimental results.

Chapter 4

Structural experiments and numerical analyses

4.1 BACKGROUND

In this chapter, we confirm the mechanical problems of practical concrete structures by experiments, apply the constitutive laws of damage mechanics summarized up to the previous chapter to finite element analysis, and introduce the analytical results compared to the experimental ones.

First, it is the reinforcing effect of fibre sheets mainly used as repair and reinforcing materials for concrete structures. Specifically, the monotonous loading and bending fracture of reinforced concrete (RC) cantilevers due to the carbon fibre sheet reinforcement, bending fracture of carbon sheet reinforced beams with 2,000,000 pre-fatigue, interfacial adhesion resistance between the concrete blocks, and reinforced with carbon fibre sheets with the same fatigue adhesion fracture. This is taken up by the cumulative damage of the attached interface due to repeated temperature loads between the concrete blocks and inorganic material with different thermal expansion coefficients.

Next, it is a matter concerning the adhesion characteristics between different materials related to the post-installed anchor fixing the ancillary equipment to the concrete structure. Here, static adhesion failure between the steel anchor, injection material, and concrete, adhesion failure due to repeated temperature load between the injection material and concrete interface with different thermal expansion coefficients; and adhesion resistance due to the difference in the roughness of the adhesion interface of the different materials are presented.

Finally, the strength characteristics related to the shear strength by a simple one-sided shear test are presented. This strength test is a test method that can be applied to the shear strength of the joint area of concrete.

4.2 BENDING FRACTURE OF REINFORCED CONCRETE CANTILEVER BEAMS WITH CARBON FIBRE SHEETS

In this chapter, we propose the carbon fibre sheet reinforcement method to reinforce the sidewalk of an existing RC pier and the passage of the vehicle

DOI: 10.1201/9781003284246-4

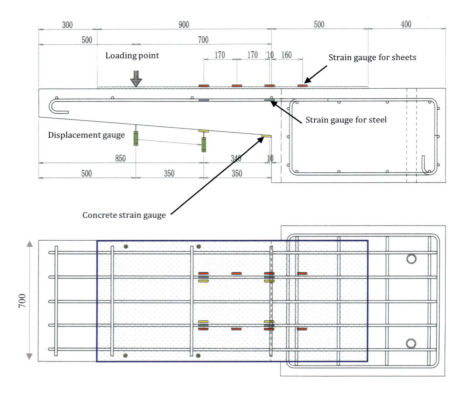

Figure 4.1 Dimension of experimental test body

and confirm the effect in the monotonous and fatigue loading of the bending fracture.

Figure 4.1 shows the test body shape. In addition, this test body is life-sized, and the material strength such as concrete also reflects the formulation at the time of construction. In this experiment, the number of carbon fibre sheets is one layer, and the orientation direction of the sheet is in the member-axis direction.

4.2.1 Characteristics of carbon fibre sheets as reinforcing materials

At present, the carbon fibre system of reinforcement of structures such as piers often adopts a method of layering a system oriented in the same direction. Therefore, we also studied the concrete structural elements which further used the carbon fibre system of one orientation. As an assumption, the anisotropic behaviour of changing the orientation of the fibres is not considered, and it was decided to treat it as an isotropic material with outstanding mechanical properties in one direction.

In general, two types of carbon fibre sheets are used to reinforce concrete structures, and although the difference between the two types is not clearly described according to the purpose of use, it has been reported that the fatigue durability of RC slabs is better in the high-strength type. In addition, as for the stress-and-strain relationships, both types exhibit elastic behaviour up to tensile strength, and show a brittle fracture that causes fibre breakage and releases stress rapidly.

The stress-and-strain relationships of carbon fibre sheets in this chapter are displayed using the elasto-plastic damage configuration equation described later, but their behaviour is elastic.

Next, the characteristics of carbon fibre sheets are briefly described. The carbon fibre is a strand in which hundreds to thousands of strands of sufficiently long monofilament with a diameter of 5–20 μm are arranged in one direction and moulded into a sheet shape, impregnated with epoxy resin, and processed into a thickness of 0.2 mm or less.

The feature is that the tensile strength is about five to ten times greater than that of the reinforcing steel bar, so that it can be attached to the concrete surface on the side subjected to tensile stress to improve the member resistance. Furthermore, by reinforcing the column to a closed carbon fibre sheet, the effect of the triaxial stress state can be expected, and a method for improving earthquake resistance performance has been devised.

Typical examples of carbon fibre-reinforced structures include bridge piers on expressways and tunnels to prevent falling.

Finally, carbon fibre sheets account for a large proportion of the two fields of bridges and buildings, and the demand has been expanding rapidly since the Hyogoken-Nanbu earthquake in 1995.

Tables 4.1 and 4.2 show the specifications of carbon fibre sheets and epoxy resins used as impregnation and adhesives.

4.2.2 Experimental results

(i) Bending fracture by monotonous loading

An experimental reinforced beam with a carbon fibre sheet is shown in Figure 4.2 before loading. Figure 4.3 shows the situation after loading. This fracture mode is where bending resistance was determined by the sheet break at the maximum bending moment point of the cantilever beam. In

Table 4.1 Specifications of carbon fibre sheet

Fibre type	Basis volume (g/m^2)	Design thickness (mm)	Tensile strength (N/mm^2)	Young Modulus (kN/mm^2)	Specific gravity
High elasticity	300	0.14	1450	668	2.1

48 Elasto-plastic damage behaviour of concrete elements

Table 4.2 Specifications of epoxy resin

Test items	Test value	Standard value
Specific gravity	1.19	1.07~1.29
Compressive strength (N/mm^2)	78	More than 70
Tensile strength (N/mm^2)	55	More than 30
Bending strength (N/mm^2)	72	More than 40
Shear strength (N/mm^2)	30	More than 10
Young Modulus (N/mm^2)	2,000	More than 1,500

Figure 4.2 Test specimen with carbon fibre sheet before loading

addition, the cracks of concrete were dispersed in several places of tensile parts in the beam, but the main crack occurred in the place where the sheet was broken.

Figure 4.4 shows the load and load point displacement curves together with the results without the sheet.

Although bending resistance can be improved by reinforcing the sheet by about 40%, it can be determined from this figure that the sheet breaks at a maximum load of 76.4 (kN) and shows a brittle fracture in which the resistance force decreases rapidly. It was also found that the load displacement curves of both were almost the same until the load (23 (kN)) at which cracks in the original test specimen occurred with or without sheets.

Furthermore, the resistance was restored after breaking the sheet with the resistance by reinforcing steel bars.

(ii) Bending fatigue by cyclic loading

In this section, in order to confirm the fatigue durability of the carbon fibre sheet RC beam for vehicle running, and to understand the mechanical characteristics after a fatigue test, a monotonous loading bending fracture experiment was conducted using the same test specimen.

Structural experiments and numerical analyses 49

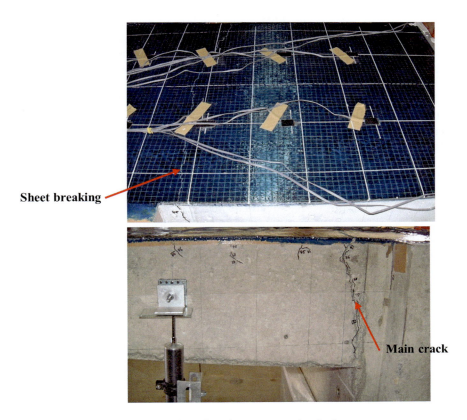

Figure 4.3 Fracture situation and cracking (monotonous loading)

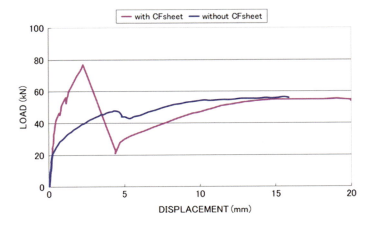

Figure 4.4 Relationship between load and displacement (monotonous loading)

50 Elasto-plastic damage behaviour of concrete elements

The fatigue load was 28 (kN) corresponding to the automobile load as a cycle of 2 (Hz), and uniformly loaded on the entire length of the beam width. In addition, the monotonous loading bending fracture experiment after pre-fatigue was carried out in the same manner as in this section (refer to Section 4.2.2(i)).

Fatigue durability of concrete members is evaluated by predicting the number of repetitions due to the main load during the service life period. In general, the number of fatigues used as one criterion for the fatigue performance of concrete structural members is used as a reference for the number of fatigues 2,000,000 times.

Therefore, this fatigue experiment also adopted the repetition frequency of 2,000,000 times by the design load.

The load-load point displacement curve of Figure 4.5 shows a tendency for plastic deformation to increase according to the number of fatigues, but the gradient of the curve is equivalent to the monotonic loading bending fracture test result of the carbon fibre sheet reinforced RC beam, and it can be said that little has changed regardless of the number of fatigues.

Since the design load of 28.3 (kN) was a load that causes cracks from the monotonous loading bending failure experiment of the carbon fibre sheet RC beam, it can be determined that further cracks have progressed due to repeated loading and damage accumulated in the beam.

Further, from Figure 4.6, cracks tended to be dispersed as in the bending fracture experiment, although the one that occurred at the maximum bending moment position of the beam was the main one.

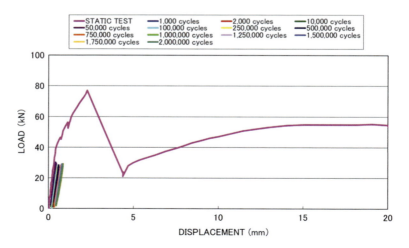

Figure 4.5 Relationship between load and displacement (cyclic loading)

Structural experiments and numerical analyses 51

Figure 4.6 Cracking (cyclic loading)

(iii) Bending fracture after pre-fatigues

In 2,000,000 repeated loading experiments, the damage evolved with the progress of cracks in concrete. However, fatigue experiments alone cannot quantitatively demonstrate how much the pre-damaged beam exhibits a decrease in bearing capacity.

Therefore, we conducted a monotonous loading bending failure experiment of the same RC beam with 2,000,000 pre-fatigue times and compared them relative to the beam without pre-fatigue to evaluate its characteristics. Figure 4.7 shows the load-to-load point displacement curves compared with the basic case. The initial value of the beam of 2,000,000 pre-fatigue is the final value of the fatigue test.

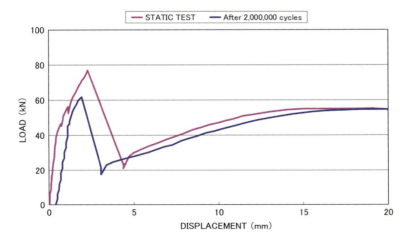

Figure 4.7 Relationship between load and displacement (after pre-fatigue)

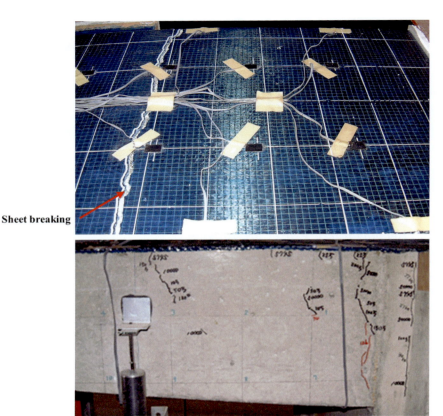

Figure 4.8 Fracture situation and cracking (after pre-fatigue)

From Figure 4.7, the bearing capacity of the beam after 2,000,000 pre-fatigue times was 61.7 (kN), which was found to be about 20% lower than that of the basic case. In addition, from Figure 4.8, it was found that the final fracture model was a carbon fibre sheet breaking at the maximum bending moment of beam as well as monotonous loading bending failure, and the behaviour after breaking tends to recover by the adhesive resistance of reinforcing steel bars and concrete as well as monotonic loading bending failure.

(iv) Pull-off test

Next, in order to confirm the adhesive resistance of carbon fibre sheets and concrete after bending failure, a pull-off test was conducted by sampling from two test specimens used in a monotonous loading fracture experiment and a monotonous loading bending failure experiment after 2,000,000 pre-fatigues.

Structural experiments and numerical analyses 53

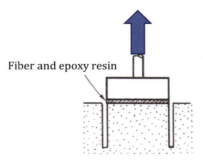

Figure 4.9 Concept of pull-off test

The pull-off test was performed by the method shown in Figure 4.9, and the tensile strength of the concrete structure surface can be directly evaluated. Furthermore, the adhesive strength of the surface material can also be measured. Here, this test was performed to confirm how much the adhesive strength between epoxy resin and concrete differs depending on the sample position after the fracture experiments.

Tests were performed by sampling from the beam after monotonic loading bending failure and after monotonic loading bending failure 2,000,000 times after pre-fatigue. Here, for the purpose of investigating the adhesive strength between the epoxy resin and concrete using the breaking position of the carbon fibre bond as a boundary, five locations were selected for a section of about 200 mm from the sheet break.

This adhesive strength is useful data for evaluating the adhesion and delaminating fracture of carbon fibre sheet and concrete, which are one of the main fracture modes that must be considered when designing carbon fibre-RC structures.

The shape of the disc to be bonded is a rectangular steel plate of 40 × 40 mm, and the maximum tensile load and the average adhesive strength are shown in Table 4.3 and Figure 4.10 indicates the positions on the specimen where the samples were selected.

Table 4.3 Results of pull-off test

Sample number	Static case		After 2,000,000 cases	
	Maximum load (kN)	Bond stress (N/mm²)	Maximum load (kN)	Bond stress (N/mm²)
①	2.75	1.72	3.82	2.39
②	3.82	2.39	4.89	3.06
③	2.91	1.82	3.66	2.29
④	4.16	2.60	4.92	3.08
⑤	5.40	3.38	5.30	3.31

Figure 4.10 Sampling points of pull-off test

Figure 4.11 shows the adhesive strength distribution of the test specimen after monotonic loading and bending failure, and that of after 2,000,000 pre-fatigues. The adhesive strength tends to increase in both test specimens corresponding to the distance from the sheet breaking position. Based on the adhesive strength at a location away from the fractured cross-section,

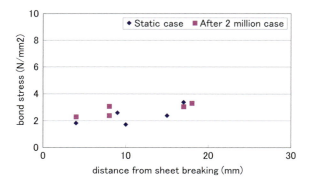

Figure 4.11 Adhesive strength distribution of the specimens

comparing the relative strength, the adhesive strength deteriorated about 30–45% near the fractured section. This suggests that damage was caused at the interface between the carbon fibre sheet and the concrete near the sheet that was broken, and that the delaminating fracture of the carbon sheet was also progressing at the same time. Further, it can be determined that there is no particularly great difference in the adhesive strength due to the presence or absence of pre-fatigue.

4.2.3 Numerical analysis by the finite element method (FEM)

(i) Identification of stress-and-strain relationships by material tests

In performing finite element analysis of a reinforced cantilever beam with a fibre sheet, the stress-and-strain relationships of reinforcing steel bars and carbon fibre sheets were identified from the elasto-plastic damage constitutive laws as well as concrete. The carbon fibre sheet was assumed to release the stress when the damage variable reached the critical value. The reinforcing bar was modelled as elasto-plastic.

Tables 4.4 and 4.5 show the material parameters in respect of fibre sheets and steel bars and Figure 4.12 shows the relationships between stress and strain and damage variables of the fibre sheet as well as concrete in tension.

(ii) Bending fracture analysis of RC beam reinforced with carbon fibre sheet

In this section, the bending fracture analysis of an RC beam reinforced with a carbon fibre sheet is described for monotonous loading.

In general, RC members are composed of composite materials of concrete and reinforcing steel bars, and when analyzing this with the finite element method, they are often modelled in two dimensions. In this case, concrete and bar are modelled and evaluated on layers with equivalent stiffness.

Table 4.4 Material parameters for constitutive laws (carbon fibre sheet)

Young's Modulus (E)	668,000 (N/mm^2)
Poisson's Ratio (ν)	0.2
Coefficient of equivalent stress (α)	0.0
Yield stress (σ_y)	1,200.0 (N/mm^2)
Coefficient of plastic hardening (K)	1,950.0 (N/mm^2)
Coefficient of plastic hardening (n)	0.0955
Coefficient of damage evolution (S$_1$)	0.055 (N/mm^2)
Coefficient of damage evolution (S$_2$)	1.755
Damage threshold (ε_{pd})	0.00
Critical damage variable (D$_{cr}$)	0.0146

56 Elasto-plastic damage behaviour of concrete elements

Table 4.5 Material parameters for constitutive laws (steel bar)

Young's Modulus (E)	200,000 (N/mm²)
Poisson's Ratio (ν)	0.3
Coefficient of equivalent stress (α)	0.0
Yield stress (σ_y)	354.0 (N/mm²)
Coefficient of plastic hardening (K)	0.12×10^{-4} (N/mm²)
Coefficient of plastic hardening (n)	0.12

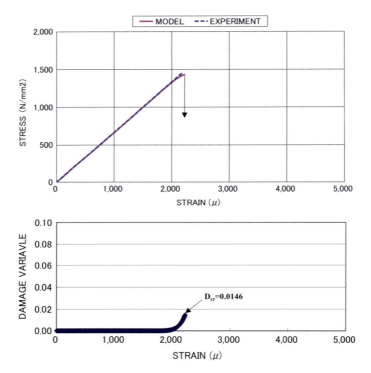

Figure 4.12 Identification of stress-and-strain relationships and damage variables for carbon fibre sheet

Similarly, the bar is a thin layer having equivalent rigidity, and the interface with the concrete is assumed to be fully fixed (displacement is continuous), and they are two-dimensional plane strain elements. Mesh division is shown in Figure 4.13.

Figure 4.14 shows a load-load point displacement curve and Figure 4.15 shows damage variables distribution that displays crack progress.

In the figure, it can be determined that the experimental results showing brittle fracture behaviour in which the load rapidly decreases is reproduced

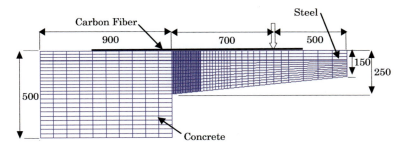

Figure 4.13

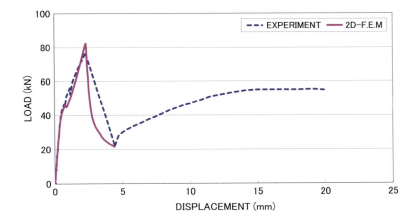

Figure 4.14 Relationship between load and displacement by analysis

well. Furthermore, the bearing capacity is 82.5 (kN) compared to the 77 (kN) of the experiment, and it corresponds well. The analysis is considered to have collapsed the structure at the time the fibre sheet is broken, and no further calculations have been performed.

From the damage variable distribution shown in Figure 4.15, not only the crack at the maximum moment position but also the concrete crack of the carbon sheet lower surface obtained a good response in the experimental results. The vertical crack is one in the experiment and two in the analysis. The vertical cracks in the experiment are located in the middle of the two vertical cracks by analysis and can be judged as a good response overall.

(iii) Bending fracture analysis of RC beam with carbon fibre sheet by pre-fatigues

The RC beam of monotonic loading and bending fracture analysis, which gave 2,000,000 pre-fatigues, was first performed using Locally Coupled

58 Elasto-plastic damage behaviour of concrete elements

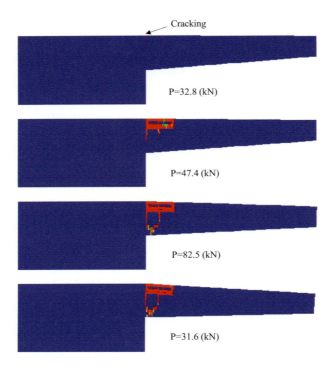

Figure 4.15 Distribution of damage variables by monotonous loading

Analysis, described in detail in Chapter 3, to calculate the damage to the concrete elements expected by 2,000,000 pre-fatigues. Next, a bending fracture analysis was conducted using the cumulative damage variable obtained by this calculation as the initial values. The procedure is shown below.

(1) Fully Coupled Analysis is performed for five cycles
(2) The stress-and-strain relationships of elements obtained by Fully Coupled Analysis are arranged
(3) Two million repeated fatigue analyses are performed using the stress-and-strain relationships as an input data for a one-dimensional fatigue model
(4) In the calculation of repeated fatigue up to 2,000,000 times, the cumulative damage variables of the focused concrete elements are arranged. That is, the cumulative damage variables of the elements exceeding the critical value Dcr is 0.99 for convenience, and the cumulative damage variables of the following elements are the calculated values
(5) Monotonous bending fracture analysis is performed using the cumulative damage variable of the concrete elements arranged in (4) as the initial value

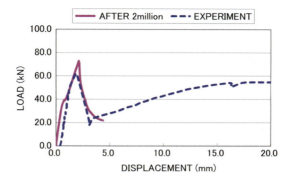

Figure 4.16 Relationship between load and displacement after 2,000,000 cycles

In the monotonic loading bending failure, the maximum loads of the analysis and the experiment were evaluated with a 9% difference, but in the bending fracture analysis after pre-fatigue shown in Figure 4.16, the maximum value of the analysis was 72.7 (kN), which was evaluated about 15% higher than the experimental value of 61.8 (kN). The reason was that the initial value is only the cumulative damage variables of the concrete, the cumulative plasticity equivalent strain of the concrete and reinforcing steel bars were not considered. It is also considered to be the limit of applying Locally Coupled Analysis to concrete structures.

Next, the load and load point displacement curves of the experimental results and the analysis are summarized and shown in Figure 4.17 for the effect of the presence or absence of pre-fatigue. The effect of 2,000,000 pre-fatigues on the bearing capacity was reduced by about 20% in the experiment, but by 12% in the analysis, the analysis estimated the decrease in

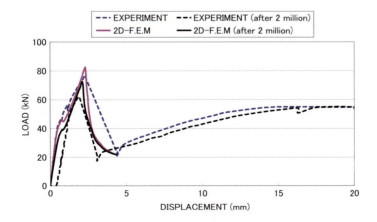

Figure 4.17 Relationship between load and displacement for all cases

60 Elasto-plastic damage behaviour of concrete elements

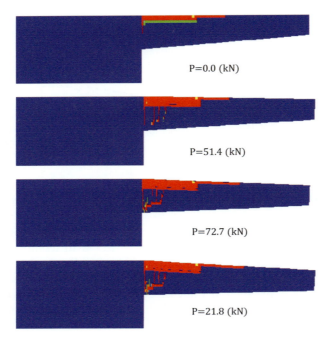

Figure 4.18 Distribution of damage variables after 2,000,000 cycles

the capacity due to the effects of pre-fatigue. However, it was found that pre-fatigue can be evaluated quantitatively by Locally Coupled Analysis, and the effect of it on the bending resistance of the overall structure can be understood analytically. In addition, the initial damage variables distribution is a good representation of the tendency to distribute macro-cracks in experiments. Further, the main crack that affects the bearing capacity occurs at the maximum bending moment point and is consistent with the carbon fibre sheet break position, and it can also be confirmed from the damage variable distribution of Figure 4.18.

4.3 ADHESION FRACTURE ANALYSES OF CARBON FIBRE SHEET REINFORCED CONCRETE

In the previous chapter, monotonous loading bending failure of an RC beam with carbon fibre sheets at real-size scale and monotonous loading bending fracture analysis after pre-fatigue are described in detail. In that analysis, it was found that the fracture of the carbon fibre sheets became a fracture mode that determined the structural strength, and in the experiment, this fracture model determined the structural strength.

On the other hand, carbon fibre-RC structural elements are determined by the mechanical properties associated with the carbon fibre sheets to improve the strength by attaching a high elasticity and high strength carbon fibre sheet. Specifically, it is called a fracture mode, and there are four items shown below.

(1) Breaking carbon fibre sheet
(2) Interfacial peeling between carbon sheet and concrete
(3) Destruction of carbon fibre anchors
(4) Delamination between the adhesive layer and the carbon fibre sheets
(5) Compressive collapse of the concrete

As for (5), it is a general RC structure destructive mode that is not directly related to the characteristics of carbon fibre sheets, but this phenomenon may induce (1), (2), and (3). In addition, the interfacial detachment between the carbon fibre sheet and concrete of these fracture modes is a typical destructive model of this reinforcement method and is still under study.

It has been pointed out that the interface peeling between the weakest concrete and the adhesive layer (concrete surface layer fracture), like the tensile brittleness of the concrete, rapidly propagates throughout the structure and greatly reduces its load resistance.

Therefore, in this chapter, we carry out monotonous loading and fatigue adhesion fracture analyses of a concrete block with carbon fibre sheet and introduce the effectiveness of the analysis method by comparing it with the experimental results on the interface peeling between carbon fibre sheet and concrete.

Specifically, the applicability of the two-dimensional elasto-plastic damage analysis to the adhesive fracture analysis of the concrete will be verified in contrast to the monotonous loading experiment results, and furthermore, the validity of the fatigue adhesion fracture analysis will be shown in contrast to the fatigue adhesion failure experiment, and the application to the design analysis code for the interface peeling mode between the concrete and the carbon fibre sheet by monotonous loading analysis and fatigue adhesion fracture will be described. The test specimens are shown in Figure 4.19 and the experimental factors are shown in Table 4.6. The concrete block has the shape of a bending test piece.

4.3.1 Adhesive fracture analysis by monotonous loading

(i) Preliminary analysis

As a result of analysing using the same elasto-plastic damage constitutive laws in Chapter 3 for adhesive fracture analysis by monotonous load, the maximum load obtained by the analysis was about 57% of the maximum

62 Elasto-plastic damage behaviour of concrete elements

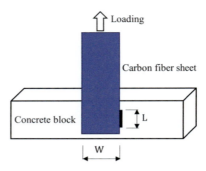

Figure 4.19 Test specimens for adhesive fracture

Table 4.6 Experimental factors and cases

CASE	Adhesive length L (mm)	Adhesive width W (mm)
CASE1	80	80
CASE2	40	80
CASE3	20	80

load of the experiment, and good correspondence results could not be obtained. Figure 4.20 shows the whole and enlarged model of the adhesive part of two-dimensional finite element analysis, Figure 4.21 shows the load and load point displacement curve using the corresponding equivalent stress of Drucker–Prager used for CASE2 in Table 4.6. The maximum value of the analysis is 5.3 (kN), compared to 9.3 (kN) of the experimental results.

Since the elasto-plastic damage analysis up to the previous chapter is limited to bending deformation close to the pure tensile state, an improved equivalent stress corresponding to the adhesion peeling behaviour between the carbon fibre sheet and the concrete, where shear deformation is predominant, was proposed and applied.

(ii) Adhesive fracture analysis by the improved equivalent stress

The improved equivalent stress was expressed in the form of constitutive laws combining Drucker–Prager's equivalent and maximum principal stress, incorporating Tresca's equivalent stress.

$$\sigma_{eq} = \alpha I_1 + \sqrt{J_2'} + \beta \sigma_{max} + \delta |\tau_{max}| \qquad (4.1)$$

σ_{eq}: Equivalent stress
I_1: The first invariant of stress

Structural experiments and numerical analyses 63

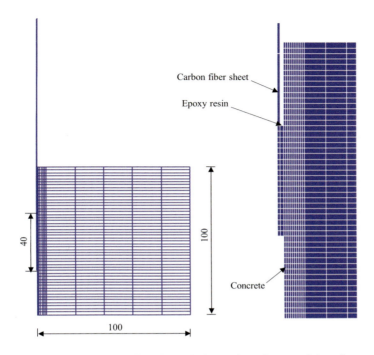

Figure 4.20 Analytical model for 2D FEM including enlarged parts of the adhesive

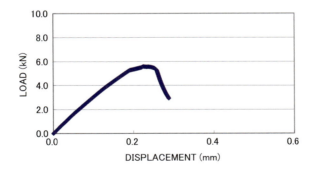

Figure 4.21 Relationship between load and displacement by Drucker–Prager's equivalent stress

J'_2: The second invariant of stress deviator
σ_{max}: Maximum principal stress
τ_{max}: Maximum shear stress
α, β, δ: Material parameters

Table 4.7 shows the material parameters of concrete. In this analysis, since the surface layer of concrete is targeted for peeling, the epoxy resin of

64 Elasto-plastic damage behaviour of concrete elements

Table 4.7 Material parameters of concrete in constitutive laws

Young's Modulus (E)	29,700 (N/mm^2)
Poisson's Ratio (ν)	0.2
Coefficient of equivalent stress (α)	0.85
Coefficient of equivalent stress (β)	0.035
Coefficient of equivalent stress (δ)	0.01
Yield stress (σ_y)	1.5 (N/mm^2)
Coefficient of plastic hardening (K)	45.0 (N/mm^2)
Coefficient of plastic hardening (n)	0.175
Coefficient of damage evolution (S$_1$)	9.5 × 10^{-7} (N/mm^2)
Coefficient of damage evolution (S$_2$)	1.55
Damage threshold (ε_{pd})	0.00
Critical damage variable (D$_{cr}$)	7.45 × 10^{-2}

Table 4.8 Material parameters of carbon fibre sheet and epoxy resin

Carbon fibre sheet	Young's Modulus (E)	350,000 (N/mm^2)
	Poisson's Ratio (ν)	0.2
Epoxy resin	Young's Modulus (E)	2,000 (N/mm^2)
	Poisson's Ratio (ν)	0.36

adhesion and the carbon fibre sheet are assumed to be elastic bodies, and the physical property values in Table 4.8 are used.

Figure 4.22 shows the relationship between the adhesion length and the maximum loads and the damage variables larger than critical value with respect to CASE3 are shown in Figure 4.23.

From Figure 4.22, it can be seen that there is no proportional relationship between the adhesive length and the maximum loads of the carbon fibre sheet. It represents the characteristic of rapidly losing its strength with the brittle fracture of concrete. In addition, from the damage variable distribution in Figure 4.23, it has been found that the surface layer peeling fracture of concrete expanded down from the upper part of the adhesive surface.

4.3.2 Adhesive fracture analysis by cyclic loading

Table 4.9 shows the material parameters of the concrete determined based on the experimental results of CASE3. In the same table, only the newly added parameters for fatigue adhesive fracture analysis are shown.

As in the previous section, fatigue failure analysis defines the damage evolution equation by dividing it into plastic fatigue damage and elastic fatigue damage, so that parameters related to damage have been added, and fatigue stress is used. That is, even within the range of elastic deformation,

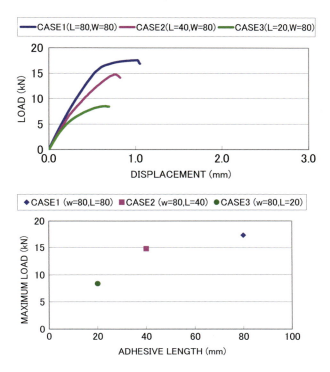

Figure 4.22 Relationship between the adhesion length and the maximum loads

if the effective equivalent stress exceeds the fatigue stress, fatigue damage develops in proportion to the corresponding elastic strain increment.

The parameters in Tables 4.8 and 4.9 were set based on the relationship between load and fatigue failure frequency obtained in the CASE3 experiment as a standard, as well as monotonic load analysis, taking into account the following assumptions and qualitative tendencies. Furthermore, the relationship between the load and the number of fatigue failures of CASE2 and CASE1 was compared in experiments and analyses, and it was confirmed that they corresponded well and the material constant was finally determined.

(1) The development of plastic damage is smaller for fatigue loads than that of for monotonous loads. Therefore, S_{pf1} is set larger than S_1
(2) The elastic damage shall be smaller than the plastic damage, and the S_{ef1} is set to be larger than the S_{pf1}
(3) The S_{pf2} shall be the same for monotonous and fatigue loads
(4) When S_{ef2} is increased, the gradient of the S-N curve displayed on the horizontal axis as the logarithm of the number of fatigue failures tends to become low

66 Elasto-plastic damage behaviour of concrete elements

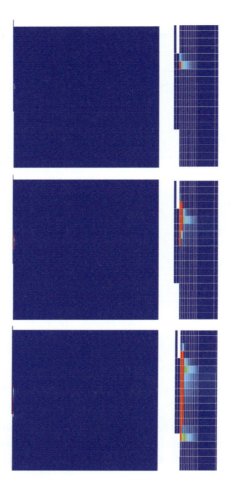

Figure 4.23 Damage variables larger than critical value for CASE3

Table 4.9 Material parameters of concrete for fatigue adhesive fracture analysis

σ_f (Fatigue stress)	1.875 (N/mm^2)
S_{pf1} (Damage parameter for plastic)	9.25×10^{-4} (N/mm^2)
S_{pf2} (Damage parameter for plastic)	1.55
S_{ef1} (Damage parameter for elastic)	4.25×10^{-3} (N/mm^2)
S_{ef2} (Damage parameter for elastic)	5.3875

Structural experiments and numerical analyses 67

Table 4.10 Fatigue adhesive experiments and analyses for CASE3

Experiments		Analyses	
Maximum loads (kN)	Fracture frequency	Maximum loads (kN)	Fracture frequency
8.836	79	8.836	141
8.653	1,404	–	–
8.575	147	–	–
8.408	879	8.408	540
8.208	744	–	–

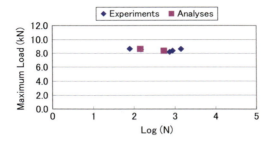

Figure 4.24 Load-fatigue fracture frequency relationship for CASE3

The repetition load is a single swinging load, and the frequency is 5 (Hz). Table 4.10 and Figure 4.24 show the experimental and analysis results of CASE3.

The number of fatigue failures in the experimental results is distributed from about 100 cycles to 1,000, except in the case of 1,404 cycles. The difference in the maximum load corresponding to the number of fatigue failures corresponds to 79 cycles and 879 is about 5%, and it can be seen that the number of fatigue failures from about 100 cycles to 1,000 occur within the range of small load fluctuations. On the other hand, the number of fatigue failures according to the analyses is 141 cycles (about 1.8 times) for 79 and 540 cycles (about 0.6 times) for 879 in the experimental results, and it is considered that the overall is almost well identified.

The results of CASE2 and CASE1 with attachment lengths of 40 mm and 80 mm are shown in Tables 4.11 and 4.12, respectively.

In CASE2, the number of fatigue times in the analysis is smaller than in the experiment but considering that the number of failures occurs from tens to hundreds of cycles within the load variation from 9.111 (kN) to 8.786 (kN), the two results correspond relatively well.

In CASE1, in order to correspond to the equivalent number of fatigue failures, the maximum loads of the analyses tend to be about 10% higher than that of experiments.

68 Elasto-plastic damage behaviour of concrete elements

Table 4.11 Fatigue adhesive experiments and analyses for CASE2

Experiments		Analyses	
Maximum loads (kN)	Fracture frequency	Maximum loads (kN)	Fracture frequency
9.310	5	–	–
9.114	56	8.932	39
8.898	332	8.898	220
8.820	3	–	–
8.786	626	8.786	528
8.649	2,376	–	–

Table 4.12 Fatigue adhesive experiments and analyses for CASE1

Experiments		Analyses	
Maximum loads (kN)	Fracture frequency	Maximum loads (kN)	Fracture frequency
8.428	7	9.355	15
7.987	572	9.006	599
7.698	13,035	–	–
6.963	93,298	–	–
6.684	256,086	–	–

Figure 4.25 shows the relationship between the load-fatigue failure frequency of CASE2 and CASE1, and Figure 4.26 shows the damage variable distribution of 39 times the number of fatigue failures of CASE2.

Equivalent stresses combined with Drucker–Prager, Tresca, and maximum principal stress for the elasto-plastic damage constitutive formula can be well evaluated for the monotonous loading and fatigue adhesive fracture strength of the concrete by prevailing shear deformation.

In addition, it was found that the maximum loads at the time of monotonous loading are not necessarily proportional to the adherence area, and the strength is rapidly lost when the surface peeling of concrete occurs. This tendency is also common to fatigue failure, and in all analysis cases, when peeling occurs, the surface layer of the entire adherent surface peels off within several fatigue times.

4.4 ACCUMULATIVE DAMAGE OF FIBRE SHEET CAUSED BY NEGATIVE THERMAL EXPANSION COEFFICIENT UNDER CYCLIC TEMPERATURE

Concrete is a standard brittle material in civil engineering. RC structures with steel bar resisting tension have been used for infrastructure such as

Structural experiments and numerical analyses 69

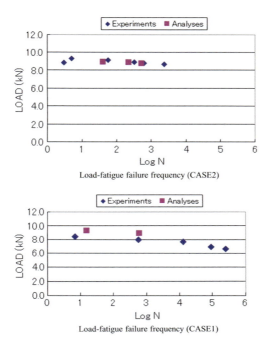

Figure 4.25 Load-fatigue fracture frequency relationship for CASE2 and CASE1

buildings, bridges, and tunnels, due to its durability and cost performance. In recent years, special attention has been paid to fibre sheets to resist against earthquake damage and prevent surface peeling fracture. The fibre sheet is of light weight (one-fifth that of steel), high strength (ten times that of steel), high rigidity, and high durability as well as easy to construct. However, the fibres have a negative thermal expansion coefficient. So, there are some concerns with mechanical interfacial properties between concrete and fibre sheets, or epoxy resin with glued fibres.

We carried out some experiments with three specimens by a freeze-thaw process test. According to them, there were accumulation of plastic strain and damage for the specimen with carbon fibre sheets by 6,000 cyclic temperature which has been changed from −10 to 3 centigrade in the winter of Northern Japan.

4.4.1 Cyclic temperature change tests

We applied the cyclic temperature change test for three specimens using a freeze-thaw process test machine. The basic shape of the specimen is a prism (100 × 100 × 400 (mm)). The properties of specimens are shown in Table 4.13. As for fibre sheet reinforcement, a method of applying a primer to a concrete surface and adhering the fibre using epoxy resin as an adhesive

70 Elasto-plastic damage behaviour of concrete elements

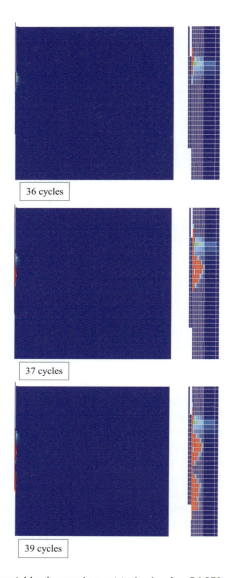

Figure 4.26 Damage variables larger than critical value for CASE2

on it is common. In this study as well, we used a specimen with one layer of fibre reinforcement using this method.

The strain gauges were respectively set on the centre and the corner of the specimens. The specimens were settled in the freeze-thaw test machine. And the temperature changes were determined considering the mean of maximum and minimum values (from −10 to 3 centigrade) in Hidaka City where the northernmost concrete piers with carbon fibre sheets were reinforced. All specimens including laminated patterns are shown in Figure 4.27.

Structural experiments and numerical analyses 71

Table 4.13 Test pieces

Series	Type	Remark
CASE1	Base concrete	$f'_{ck} = 30 (N/mm^2)$
CASE2	Primer	T = 0.3 (mm)
CASE3	Carbon fibre	T = 0.22 (mm) $\alpha = -1.0 \times 10^{-6} (/°C)$

f'_{ck}: Design strength of concrete specimen

t: Thickness, α means the thermal expansion coefficient

Thickness of CASE3 includes epoxy resin as glue.

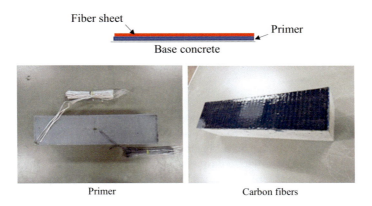

Figure 4.27 Specimens of fibre sheets reinforcement

There are generally two ways to set the specimens in air or in water. If we set them in water, we couldn't specify the deteriorated factors caused by the different thermal expansion coefficient of fibres or the frost damage by water. Therefore, we carried out the experiments simply in air in order to be sure of the effect by the different thermal expansion coefficient of fibres. In addition, it has been known that the diffusional coefficient of concrete in the wet has strong nonlinearity around zero centigrade. But it is very important and difficult to measure and evaluate the nonlinearity on the surface of concrete. And most of the structures in real environmental conditions haven't been exposed in the wet at all. So, in these experiments, we focused on the properties of fibres.

4.4.2 Relationship between temperature and strains for specimens

Figure 4.28 shows the temperature change of one cycle of the freeze-thaw process, which was changed regularly considering the maximum and minimum range of the machine's capacity. In this study, the freeze-thaw test was conducted in air conditions and the relationship between temperature and strains that are gauged the centre of specimen surface are shown in this figure.

72 Elasto-plastic damage behaviour of concrete elements

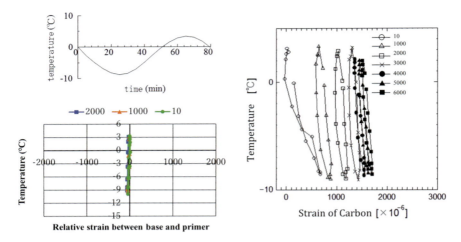

Figure 4.28 Temperature change and the results of strain curves

Judging from the relative strain between the Base and Primer, up to 2,000 cycles, they are fully bonded to each other, and its distribution is also almost near zero compared to the distribution of carbon fibres. Therefore, it can be determined that the cumulative damage between the Base and Primer has not occurred. On the other hand, since the strains of the carbon fibres increase due to the repetitive temperature load, the strain history of that is focused on.

The strains of the carbon fibres have been changed due to cyclic numbers. In particular, the gradients of the loop curves have been increased under cyclic temperature. The inverse of the gradient has shown the apparent thermal expansion coefficient of the specimens including epoxy resin. And then these phenomena meant that fibres have split out of the epoxy resin. Therefore, we considered that the damage has also been accumulated between the epoxy resin and carbon fibres.

Damage between the carbon fibre and the epoxy resin due to repeated temperature loads is evaluated with the two indicators shown in Figure 4.29. One is a method of approximating the history curve at the time of the corresponding repetition cycle with an ellipse and evaluating the area as strain energy. Another method is to use the gradient of the long axis of an approximate elliptic curve. The values of the strain energy of carbon fibre have decreased due to cyclic numbers. This means that the absorbable ability of materials corresponding to deformation has been decreased.

Secondly, we will focus on the gradient of the straight line obtained from the major axis of the ellipse approximating the history curves. Generally speaking, there is linear relationship between the thermal strain increment and temperature increment using the thermal expansion coefficient:

$$\Delta\varepsilon = \alpha\Delta T \qquad (4.2)$$

Structural experiments and numerical analyses 73

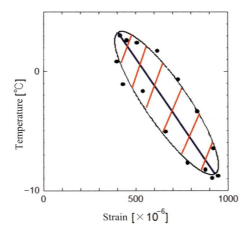

Figure 4.29 Two indicators for assessing cumulative damage of carbon fibre

where, $\Delta\varepsilon$ is thermal strain increment, α is thermal expansion coefficient, ΔT is temperature increment. The thermal expansion coefficients of carbon fibre have been decreased due to cyclic numbers. In this research, the temperature increment ΔT has been constant from –10 to 3 degrees. But the thermal strain increment $\Delta\varepsilon$ has been decreased. This means that the thermal expansion coefficients α have deteriorated by accumulative damage D that is the damage variable. Therefore, the thermal expansion coefficients α have been changed:

$\alpha(1-D)$. The damage has accumulated between fibres and epoxy resin that glued the fibres and primer.

Figure 4.30 shows the relationship between strain energy and the apparent thermal expansion coefficient for the number of repeated temperature loads. Since both parameters decrease in response to the increase in the number of repetitions, it can be judged that it can be used as an index to evaluate cumulative damage.

4.5 ADHESIVE CHARACTERISTICS BETWEEN INORGANIC INJECTION MATERIALS AND CONCRETE IN THE POST-INSTALLED ANCHOR METHOD

We have used organic material like epoxy and urethane resin to glue steel or plastic materials to concrete structures. In particular, epoxy resin is useful for cracking repair or injectable material. But these materials have some flaws, for example, deterioration by UV light, nonconformity under water construction, and non-filling in case of the upward injection in holes.

74 Elasto-plastic damage behaviour of concrete elements

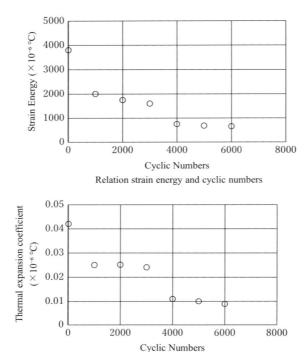

Figure 4.30 Relationship between two indicators and cyclic numbers

We have tried to adopt the inorganic material which could make up for defects as the injectable anchor material to connect to a steel anchor and concrete structures in cases of thin concrete structures.

Adhesive force has been easily defined by the interfacial area of two materials or hypothesis of shear or diagonal fracture. But we have to consider fracture modes that define interfacial material, shear of injectable material, and bond-splitting fracture of concrete.

In this section, some experiments are conducted to clear the adhesive force between the steel anchor and injectable material, or the injectable material and concrete block considering fracture modes by the punching-out shearing test. In addition, the experimental results are shown by the wedge-shaped injectable form compared with those of the normal cylindrical cases.

In this research, we use cylindrical concrete specimens (outer diameter = 150 mm). And the compressive strength of concrete is 50 (N/mm^2) and 55 (N/mm^2) for injectable material.

One of the specimens and the concept of testing method are shown in Figure 4.31. And we adopt the displacement control method by the Amsler

Structural experiments and numerical analyses 75

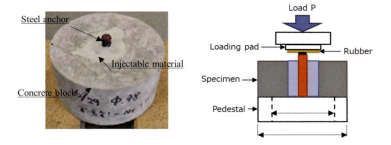

Figure 4.31 Specimen and loading method

that was controlled at a constant speed (2.0 (mm/min)). The injectable length that is equal to the specimen height is adjusted: 30, 40, 55, 65, and 80 (mm). And the injectable diameter is also adjusted: 48, 60, and 76 (mm). The steel anchor protrusion length from the top or bottom surface of specimen is 20 (mm).

Some previous studies were carried out by the pulling-off test which needed special steel jigs and tools. Generally, the mechanism of adhesion is defined by shear behaviour within a material or an interfacial border between two materials. In cases of isotropic material, we could apply the punching-out test to this research from a theoretical point of view. We found out that there were almost the same fracture modes in pulling-off or punching-out.

And we narrow down the concrete structure to about 100 (mm) thickness. These are the main differences between previous studies and ours.

4.5.1 Influence of injectable diameter and length

In this experiment, to improve the adhesion between the injection material and the concrete, the surface treatment that is usually performed at the construction site on the surface of the concrete to be injected is performed with a brush.

Two main fracture modes are found whether the injectable length was less than 55 (mm) or not. One is a bond-splitting fracture under 55 (mm) in length. And then, the other case of more than 65 (mm) in length is found not only in bond-splitting but also in the cone-shaped fracture. The maximum loads due to the influence of injectable diameter and length are shown in Figure 4.32 corresponding to two fracture modes.

The maximum load tended to be in inverse proportion to the injectable diameter in cases of more than 65 (mm) in length. We thought the height of the cone shape became lower in inverse proportion to the injectable diameter, and then the total area against the load got smaller.

76 Elasto-plastic damage behaviour of concrete elements

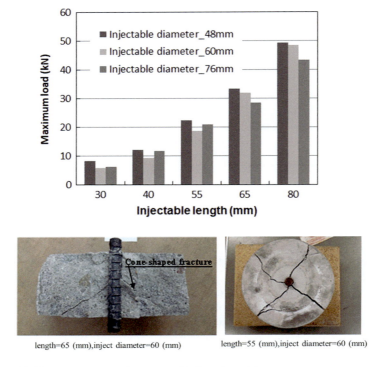

Figure 4.32 The maximum loads—injectable length and diameters

4.5.2 Improvement of adhesive force by the wedge-shape effect

Post-installed anchors have also been applied to many civil engineering and building structures. The injection material and the concrete surface are subjected to surface treatment using a brush or the like to improve the adhesion strength. It has been pointed out that the surface treatment of numerous injection holes increases the construction cost, and that sufficient adhesion performance cannot be ensured at construction sites where surface treatment cannot be performed. To cope with this, the shape of the injection hole is made into a wedge shape and the effect is confirmed.

The parameters of experiments are shown in Table 4.14.

(i) Adhesion characteristics of injection material and concrete

First, the adhesion characteristics between the injection material and the concrete were investigated by changing the inclination angle.

Table 4.14 Parameters of specimens

	CASE1	CASE2	CASE3
Design strength of concrete	30 (N/mm^2)	40 (N/mm^2)	50 (N/mm^2)
Inclination angle of wedge-shape (α)	3,5,7 degrees	3,5,7 degrees	3,5,7 degrees
Upper diameter of injectable form	45 (mm)	45 (mm)	45 (mm)
Injectable length	70 (mm)	70 (mm)	70 (mm)

The steel anchor uses the SD345 of the D13.

The relationship between the maximum load and the concrete design reference strength and inclination angle is shown in Figure 4.33.

The design strength of concrete has little influence on the adhesive force. However, the inclination angles have a positive effect on that. The bigger the inclination angle becomes, the bigger the adhesive force becomes. That maximum load of 5 or 7 degrees is more than double the normal cylindrical form. And then in cases of the design strength of concrete of 40 (N/mm^2) and 50 (N/mm^2), the maximum loads are almost the same. We thought there was a certain effective inclination angle on the force.

The fracture modes are interfacial in cases of normal cylindrical form. But the mode is shifted into bond-splitting and a cone shape due to inclination angles. There are some improved effects on the adhesive resistant force by wedge angles. Two typical modes are shown in Figure 4.34. When the design reference strength is 30 (N/mm^2) and the inclination angle is 7

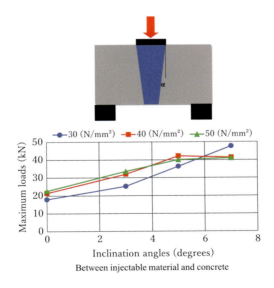

Figure 4.33 The maximum loads—inclination angles in case of loading on injectable material

78 Elasto-plastic damage behaviour of concrete elements

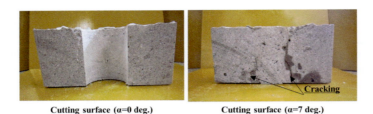

Figure 4.34 Differences in fracture modes due to the inclination angles

degrees, it can be estimated from the failure mode that the cone-like shear resistance of the concrete in addition to the slip resistance at the interface more affected the maximum load.

(ii) Adhesion characteristics of anchor and injection material and concrete

Figure 4.35 shows the relationship between the maximum loads and the inclination angles. The influence of the design reference strength of concrete on the maximum load is less than in the previous case. From this, it can be said that the wedge effect can be expected regardless of the strength of the base concrete. In addition, it is found that the influence

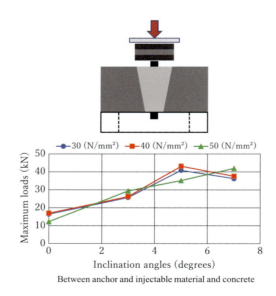

Figure 4.35 The maximum loads—inclination angles in case of loading on steel anchor

of the inclination angles is the same as in the previous case, and as the angle increases, the maximum load increases, and 5–7 degrees is the optimal angle.

Next, the feature of the fracture mode is described as an example of the design reference strength of 40 (N/mm^2). In a normal cylindrical injection hole, having an inclination angle of 0 degrees, the interface fracture mode between the injection material and concrete can be seen, and in the case of 3 degrees or more, bearing pressure fracture by the node of the anchor precedes and is subsequently a cone-shaped fracture, which is the shear fracture of the injection material, and then it is found that it progresses to an interface fracture between the injection material and concrete. Further, as the inclination angle becomes larger, the bearing pressure fracture section due to the nodes of the anchor tends to be longer (refer to Figure 4.36).

On the bottom surface of the specimen, as in the previous cases, the bond-splitting fracture can be seen radiating from the anchor.

Finally, regarding the wedge effect, it is found that setting the inclination angle from 5 to 7 degrees yielded an adhesive resistance equivalent to that of a case where the concrete surface of the normal injection hole was treated with a brush.

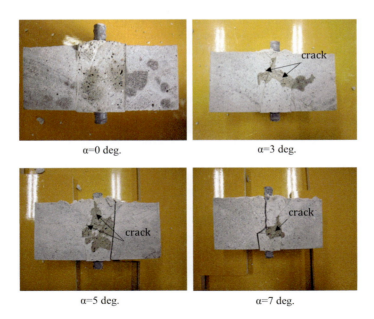

Figure 4.36 Failure mode due to different wedge angles

4.5.3 Numerical analyses by return mapping algorithm

In this section, numerical analysis results using the two-dimensional finite element method are presented. In this analysis, the implicit algorithm called the return mapping method is used. The main features of the return mapping method are shown below:

(1) Using the strain increment and the increment of the plastic multiplier in the time interval $[t_n, t_{n+1}]$, the elastic strain at t_{n+1} and the internal variable are obtained
(2) Depending on the condition of the incremental value of the plastic multiplier, the evaluation is performed by moving to the elastic trial step and the plastic corrector step and performing iterative calculations and confirming that the stress at t_{n+1} converges on the yield surface
(3) A tangential stiffness tensor that relates the stress and strain used at this time is required when deriving the consistent tangential operator in this algorithm

The elasto-plastic damage tensor introduced in Chapters 2 and 3 represents tangential stiffness and is easily applied to the return mapping method.

Here, as a two-dimensional axisymmetric problem, the adhesive characteristics between the injection material and concrete, and between steel anchor and the injection material and concrete are analysed.

The feature of these analyses is that when the values of the equivalent plastic stress including damage obtained from the constitutive laws and the equivalent plastic strain are directly input and the stress is not released abruptly, it is possible to evaluate that the stress converges on the yield surface even in the strain-softening region. The analyses are a half model in consideration of symmetry.

The following analysis results show a case with a concrete design reference strength of 40 (N/mm$^{2)}$ with an inclination angle of 5 degrees.

Figure 4.37 shows the relationship between the equivalent plastic stress and the equivalent plastic strain of the concrete, the injectable material, and the anchor. Since the fracture mode is confirmed and there is little difference in the mechanical characteristics of the injectable material and the concrete, the equivalent plastic stress and the equivalent plastic strain of the concrete and the injection material are the same value.

Figure 4.38 shows the adhesive fracture between the injection material and the concrete. As in the experiments, the failure at the interface between the two can be reproduced, but the shear failure of the concrete on the underside cannot be evaluated.

Figure 4.39 shows the adhesion fracture between the anchor, the injectable material, and the concrete. The bearing pressure fracture from the

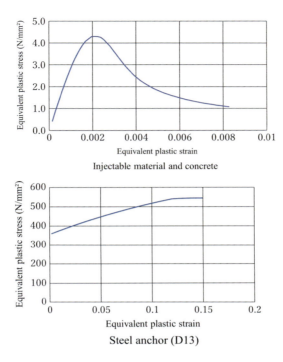

Figure 4.37 Equivalent plastic stress and strain of materials

nodal tip of the anchor progresses from the top surface to the lower surface. In this analysis, it is not possible to evaluate the progress from the bearing fracture to the shear failure of the injection material, or to the interface failure of the injection material and concrete.

It is necessary to confirm the failure mode by multi-scale analysis that models constituent materials such as aggregates and constitutive laws limited to adhesive characteristics at interfaces of different materials including surface treatment.

4.6 CUMULATIVE DAMAGE TO INTERFACES DUE TO REPEATED TEMPERATURE LOADS OF TWO INORGANIC MATERIALS WITH DIFFERENT THERMAL EXPANSION COEFFICIENTS

The inorganic injection material used in the previous section has a thermal expansion coefficient 1.5 times that of concrete. When subjected to repeated temperature loads, this feature causes cumulative damage at the interface of both materials, and there is a concern that it eventually causes interfacial delamination.

82 Elasto-plastic damage behaviour of concrete elements

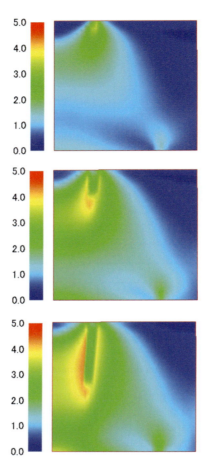

Figure 4.38 Distribution of equivalent plastic stress between injectable material and concrete (unit: N/mm²)

Therefore, in this section, the cumulative damage at the interface is evaluated experimentally by repeatedly applying a temperature load to the specimen using two cases in which the adhesive interface of both materials is surface treated and when it is not surface treated.

4.6.1 Experimental method

A prismatic column specimen of concrete with a compressive strength of 40 (N/mm²) and an injection material of 50 (N/mm²) is prepared, and the specific survey method is to attach a strain gauge to the central part of each specimen (refer to Section 4.4.2). A repeating temperature load is applied to the specimen using a freeze-and-thaw test machine, and the relative

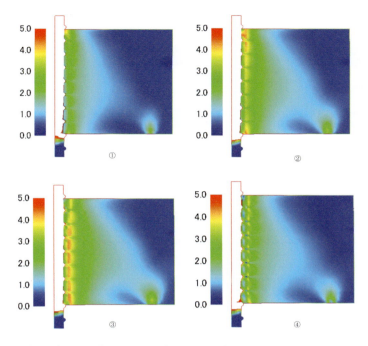

Figure 4.39 Distribution of equivalent plastic stress between steel anchor and injectable material and concrete (unit: N/mm²)

difference of strain obtained from the strain gauge attached to each material is calculated. Next, the relative strain curve obtained from a single material and the relative strain curve obtained from a specimen in which both materials are bonded are approximated to an ellipse, and the distance between the two is calculated corresponding to the number of repetitions.

Figure 4.40 shows a measurement method of relative strain and an example of measuring relative strain. It is based on the relative strain curve of a single material.

4.6.2 Experimental results

When damage develops at the adhesive interface, the effective adhesive area decreases. This can be assumed that when the delamination of the interface progresses, the relative strain curve of the adhered specimen is equivalent to the phenomenon of approaching the reference curve. From this, it is determined that the distance between the two strain curves calculated and corresponds to the damage variables.

Figure 4.41 shows the damage variable-repetition frequency and an example of an interfacial detached fracture. The damage variables when the surface treatment is not performed evolves linearly, whereas the damage

84 Elasto-plastic damage behaviour of concrete elements

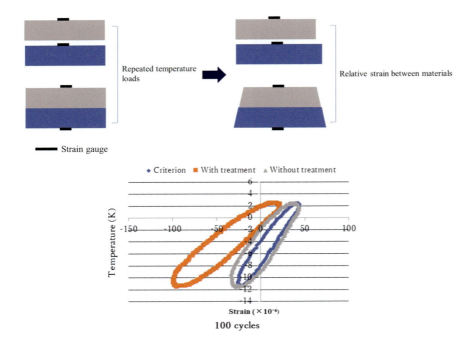

Figure 4.40 Relative strain and measurement example

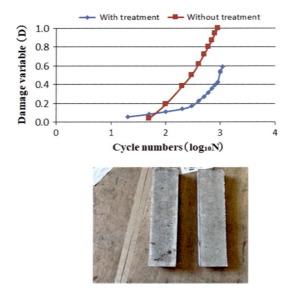

Figure 4.41 Cumulative damage variable and repetitive temperature load numbers

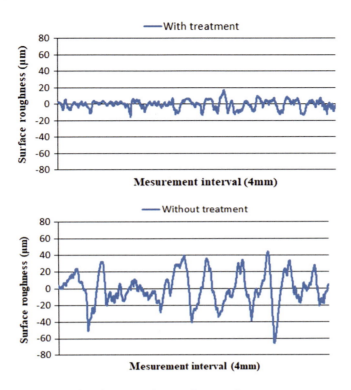

Figure 4.42 Roughness distribution with or without surface treatment

variable, when performed, develops exponentially and the adhesion performance is also improved. In the figure, a state in which the specimen without surface treatment finally caused interface separation is also presented.

4.6.3 Effects of roughness

When the concrete surface of the adhesive surface is treated with a brush and when it is not, a difference of about two to three times the adhesive resistance force can be seen.

Figure 4.42 shows the difference depending on the presence or absence of surface treatment measured with a roughness meter. The roughness is greater when surface treatment is not performed, and the roughness of the one which has been carried out is small and average. The results of studies on inorganic and metallic materials have also shown that surfaces with average and moderate roughness improve adhesion performance.

In order to evaluate the adhesion performance of the interface considering roughness, multi-scale analysis represented by the homogenization method seems to be effective. There is also a method of analyzing the roughness distribution curve to evaluate the difference in attachment intensity from the probability of expression of the measurement interval of the principal component waves.

4.7 SHEAR STRENGTH OF CONCRETE

The strength of concrete is defined by compression, bending, and tensile testing respectively. However, regarding shear strength, several test methods have been proposed, but have not yet been established. It is thought that one of the causes is that shear strength is rarely utilized in the design of concrete structures.

However, as the scale of a concrete structure increases, it is necessary to provide several construction joints in the vertical direction. The strength of these joints has a property that decreases more than other parts, and some treatment agents that improve this property are applied. In order to evaluate the improving effect including the strength of the treatment agents on the joint surface, shear testing is the optimal method.

Therefore, in this section, focusing on the direct shear test method of concrete, we introduce analysis examples that verify its usefulness experimentally, and evaluate the cracking progression.

4.7.1 Experiments

The shear strength test is performed as shown in Figure 4.43. The main feature of the specimen is that steps are provided above and below the left and right specimens at the boundary of the shear section, and compression forces are applied from the upper surface of one specimen to directly investigate the shear resistance forces.

This time, a method in which a pre-crack is provided in the shear section and an improved method is adopted so that cracks other than the shear section do not occur. Since steel plates are installed on both sides to hold the specimen in the vertical direction, friction occurs between the specimen and the plate. To reduce this, Teflon sheets are installed in pairs.

The concrete used for the test was studied with a design reference strength of 30 (N/mm^2) and a shear length of 50, 100, and 150 (mm). Figure 4.43 shows the state of the fracture after the test.

4.7.2 Analyses

The analysis covers a case with a shear section length of 50 (mm) as a two-dimensional plane strain and ignores the effect of friction on both sides.

As in the previous section, the analysis method is a return mapping method, and for the main purpose of how accurately the rapid strain softening due to brittleness can be analysed, the equivalent plastic stress-equivalent plastic strain curve is changed in two cases.

Structural experiments and numerical analyses 87

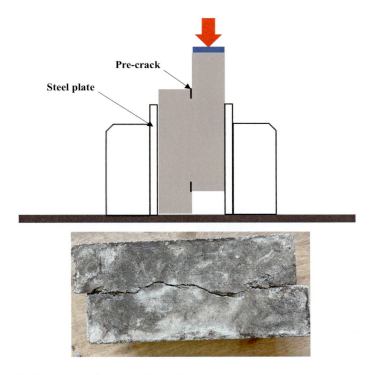

Figure 4.43 Shear strength test method of concrete

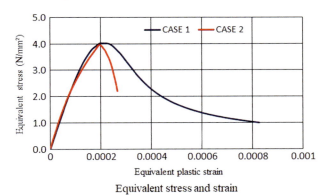

Figure 4.44 Relationship between equivalent plastic stresses and strains

Figure 4.44 shows the relationship between the equivalent plastic stresses and the equivalent plastic strains in the two cases analyzed. CASE2 is a distribution case that assumes more brittle failure, and this case is the limit that satisfies the convergence conditions at the time of calculation.

88 Elasto-plastic damage behaviour of concrete elements

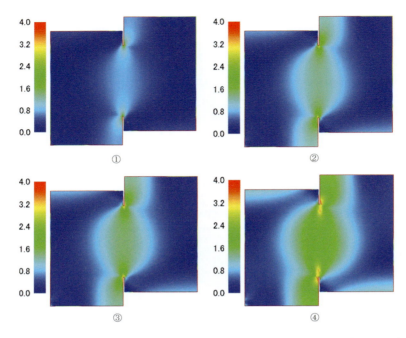

Figure 4.45 Distribution of equivalent plastic stress by Tresca's yield function (unit: N/mm^2)

Figure 4.45 shows the equivalent plastic stresses distribution of CASE2 analyzed using the Tresca yield function. It is consistent with the experimental results in which stress concentration occurs at the tip of the pre-crack, from which the crack develops along the shear section.

Chapter 5

Applicability of damage mechanics to the concrete field

5.1 BACKGROUND

It is common for concrete structural elements to be designed and constructed as members of several metres or more using the representative volume elements of the concrete with a minimum side of about 100 (mm) and the strength of each side such as compression, tension, bending, etc., is confirmed by experiments. Of course, it is more efficient to estimate each design strength from the design reference strength.

Furthermore, as a structural member, if it is the same material, it is often designed as a composite structure with a rebar of constant strength, elastic modulus, and the like. This is because, regardless of the concrete mix proportion, the concrete has a ratio of 10 or more between compressive strength and tensile strength, and the rebar is arranged at a place where tensile stress occurs in the element, to avoid brittle failure as a member, and to ensure load-bearing capacity and toughness as a structure.

At this time, to understand the stress state of each part from the serviceability limit state to the ultimate limit state against all the working loads, the stress-and-strain relationship of each material, called the constitutive laws, are important.

Next, it has been pointed out that the constructed concrete structural members deteriorate chemically due to carbonation, salt damage, etc., and the initial performance deteriorates over time, making it impossible to maintain durability and load resistance.

In response to these issues, various repair and reinforcement technologies are applied according to the situation of deterioration in performance, and the life of the structure is extended.

As described above, the concrete structural elements are designed and constructed as beams, columns, and planar members based on the mechanical characteristics of the constituent materials mainly composed of concrete and rebar. After that, repairs and reinforcements are carried out according to ageing deterioration, and it is maintained to retain the performance that meets the initial management standards for a useful life.

DOI: 10.1201/9781003284246-5

The applicability of damage mechanics to the field of concrete is described from the following three perspectives based on the above sequence of events.

5.2 APPLICABILITY OF DAMAGE MECHANICS

5.2.1 Representative volume elements that define material strength and constitutive laws

The strength of the concrete generally represents the compressive strength. It is said that it is possible to estimate the strength of tension, bending, etc., from this strength.

The concrete is composed of four main materials: Water, cement, fine aggregate, and coarse aggregate, and is formed from different materials in a wide range of granular shapes, from coarse aggregates with a grain shape of about 25 (mm) to cement with a grain shape of 0.1 (mm) or less.

The dimensions for the strength test of compression and tensile strength are specified corresponding to the grain shape of the coarse aggregate, and a cylindrical specimen of Φ100 (mm) and a height of 200 (mm) is adopted at 25 (mm) or less, and a prismatic specimen of 100 × (mm), 100 (mm) × 400 (mm) is adopted by Japan Industrial Standard for the bending strength.

The compressive strength of the concrete is generally proportional to the mass ratio of cement and water, and in ordinary concretes with a compressive strength of 40–50 (N/mm^2) or less, fine cracks generated in the mortar matrix, having a smaller strength than that of coarse aggregate, grow and coalesce to cause material failure.

On the other hand, in a concrete called high-strength concrete of 60 (N/mm^2) or more, the strength of the coarse aggregate may be determined more than the mortar matrix, and the cracks generated in the coarse aggregate and the mortar matrix will grow and coalesce, and eventually cause material failure.

As described above, the process of occurrence, growth, and coalescence of cracks between the constituent materials differs depending on the magnitude of the compressive strength of concrete.

In addition, in the stress-strain curve, regardless of the magnitude of the compressive strength, the corresponding strain tends to be within a certain range. However, strain-softening behaviour after the maximum stress becomes more rapid in a concrete with high strength. That is, the high-strength concrete exhibits a more brittle fracture.

The stress-and-strain relationship, including each strength such as compression, tension, and bending, is defined by the fracture associated with cracking, growth, and coalescence between materials in representative volume elements.

It is necessary to analyze the stress-and-strain relationship, including the rapid strain softening of the compressive behaviour of high-strength concrete, to the extent that coarse aggregates and mortar matrices can be minimally modelled.

It can be considered as an analysis of the mesoregion targeting representative volume elements. In this case, damage mechanics, which expresses the occurrence, growth, and coalescence of cracks between materials and the materials themselves as damage variables, which are the amount of states, is a mechanical system that can be applied to such analyses. It should also be noted that this mechanics is based on conventional elasto-plastic mechanics and can be applied to various numerical analysis methods, including the finite element method.

5.2.2 Structural members that define structural performance

It falls into the category of structural members with a scale of several metres or more, and some structural elements, but except for concrete lining in mountain tunnels and gravity dams, most structural members are designed and constructed as reinforced concrete. Therefore, it is necessary to understand the adhesion characteristics of concrete and rebar in structural members or representative volume elements.

The design of concrete structures in Japan currently employs a limit state design method, which basically examines the performance of each of the serviceability limit states and the ultimate limit state so as to satisfy each performance.

Specifically, in the serviceability limit state, the stress of the rebar is less than or equal to the specified value, and the bending crack width of the concrete surface is less than or equal to the allowable value defined from the surrounding environmental conditions. In addition, in the ultimate limit state, in the stress-and-strain relationship of the material for design, the bending bearing strength, shear bearing force, and axial bearing force, when the ultimate strain of the concrete is reached, are calculated and the safety performance is confirmed by comparing them with the maximum cross-sectional force generated by the assumed external force.

The allowable bending crack width is defined as 0.5 (mm) or less even in a general environment, and to calculate this crack width it is necessary to evaluate not only the current experimental formula with limited conditions but also numerical analyses.

At this structural member level, it is a prerequisite of the design that the concrete and the rebar work together to resist external forces. However, when the structural elements reach not only the serviceability limit state that allows the crack width but also the ultimate limit state, the adhesion between the rebar and the concrete is reduced, including a partial section.

Usually, deformed steel bars are used for rebar, and protrusions called nodes of about 1 (mm) are arranged at intervals of more than a dozen millimetres. The bearing resistance of the mortar part of this node spacing accounts for most of the mechanical adhesive and plays an important role as an adhesion between the concrete and the rebar.

To consider the influence of crack widths of 0.5 (mm) or less and rebar nodes of about 1 (mm), it is essential to conduct numerical analyses of the model that takes into account the mesoregions at the structural member level similar to the representative volume element level, and to evaluate the process of structural collapse through growth and coalescence of cracks. In this performance evaluation at the structural member level, since cracks are an index, the applicability of damage mechanics that can systematically handle this in the development equation of the damage variable is high.

5.2.3 Time to define performance maintenance

Generally, concrete structural members are designed with a design service life set around 50 years. If the performance deteriorates within this period, repair and reinforcement are carried out based on the set management standards as a guideline, and maintenance and management are performed in anticipation of the performance degradation from the beginning.

Ageing deterioration of the member is often caused by mass transfer due to chemical reactions such as salt damage, carbonation, and alkaline aggregate reaction. Among these, deterioration in structural performance due to internal rebar corrosion is more problematic than the deterioration of the concrete.

First, as the main deterioration mechanism common to all structural members, the movement and penetration of water, oxygen, carbon dioxide, chloride ions, etc., into the concrete is a direct factor in steel corrosion, and the influence of temperature and humidity is also added, and the distance from the concrete surface of the steel materials, the crack width, and depth have a great influence on the corrosion rate.

In particular, salt damage and carbonation cause the volumetric expansion of the steel materials arranged in the concrete due to corrosion, this causes cracks in the concrete from the steel side, and a part of the concrete to be missing. In the end, it is a common deterioration process that causes loss of the effective area of the steel material itself, loss of member bearing capacity, and it cannot resist external forces, so it is impossible to investigate this process from a normal visual inspection.

Furthermore, before deterioration becomes apparent, it is often from 20 to 30 years or more, and it is difficult to clearly identify factors related to deterioration and the influence between factors, and to accurately predict the deterioration process.

Assuming a situation in which chemical deterioration also proceeds at the same time while being repeatedly damaged by external forces as a structural member such as fatigue, the proposition in this section is how to elucidate this dual problem.

Of course, it is also necessary to consider responding to natural disasters such as earthquakes and floods, as well as changes in conditions due to man-made disasters such as fires and accidents.

The deterioration of steel materials due to chemical reactions requires elucidation of the following major issues in order to understand the mechanical properties over time.

(i) Chemical reactions due to mass transfer cannot be directly associated with physical quantities to strains, etc.
(ii) Changes in the volumetric expansion, modulus of elasticity, and yield strength of steel materials due to corrosion cannot be ascertained
(iii) Failure to assess the deterioration of the adhesion properties of corroded steel materials and concrete

However, the reduction in the effective area of the steel materials due to corrosion can be defined by the damage variables. Furthermore, the curve representing the process of performance degradation due to ageing is very similar to the curve representing the amount of state considering the development of the damage variables.

Taking these points into account, the effective area due to corrosion of the steel materials and the decrease in the adhesion strength with the concrete can be evaluated by damage mechanics.

The next most important item is that material deterioration over time also has a significant impact on performance evaluation after repair and reinforcement. In particular, when the performance is improved by adhering the newly constructed materials to the existing members, it is expected that the fracture of the interface with the existing surface becomes a fracture mode rather than a fracture of the material including the newly constructed part and is different from the required performance.

In addition, depending on the methods of repair/reinforcement, the resistance will be improved, but it will shift to a rapid brittle failure, and the final fracture mode may also have a different case from the initial design, so it is necessary to verify the performance after repair/reinforcement.

At present, damage mechanics can only be evaluated in a limited way by reducing the effective area of steel materials in the process of material deterioration over time after repair and reinforcement, but it can be applied to the effect of improving the resistance after repair and reinforcement, confirmation of the fracture mode, etc. Furthermore, the ability to evaluate the effects of repetitive loads after repair and reinforcement using cumulative damage variables is a notable advantage compared to other dynamic systems.

Finally, the common item among the above three perspectives is the crack. However, since the target period spans from several months to several decades after construction, a model that can be evaluated uniformly has not been established. The main factor is that mechanical physical quantities such as strains and stresses are not organically related to physical quantities defined by chemical reactions due to mass transfer.

However, representative volume elements and structural member elements related to strength and mechanical performance are considered to be compatible if they are modelled with a unified dynamic system and are easier to connect to the performance maintenance evaluation strategy.

In recent years, discontinuous mechanics has attracted attention as one of the systems that can model aggregates and mortar matrices of granular bodies, and evaluate the break of each material and the detachment between materials. This is an effective system that can express the formation of discontinuous surfaces between materials, but sufficient results have not yet been obtained to calculate the crack width.

However, a method for combining the relationship between microcrack density and damage variables, a smeared crack model, and the like have been proposed, and it has been suggested to the stage where crack generation, coalescence, and progress can be evaluated with high accuracy.

Reference books

D.R.J. Owen, E. Hinton, *Finite Elements In Plasticity: Theory and Practice*, Kagaku-Gijutsu Publishing Co., Inc.
L.M. Kachanov, *Introduction to Continuum Damage Mechanics*, Martinus Nijhoff Publishers
J. Lemaitre, *A Course on Damage Mechanics*, Springer
J. Lemaitre, R. Desmorat, *Engineering Damage Mechanics*, Springer
S. Murakami, *Continuum Damage Mechanics*, Morikita Shuppan Co., Inc.
J. Lemaitre, J.L. Chaboche, *Mechanics of Solid Materials*, Cambridge University Press
D. Krajcinovic, *Damage Mechanics*, Elsevier
J.C. Simo, T.J.R. Hughes, *Computational Inelasticity*, Springer
Y.Toi, *A Course on Computational Solid Mechanics – Modeling and Simulation of Materials and Structures*, Corona Publishing Co., Ltd
E.A. de Souza Neto, D. Peric, D.R.J. Owen, *Computational Methods for Plasticity: Theory and Applications*, Morikita Shuppan Co., Inc.
Japan Society of Civil Engineers, *Standard Specifications For Concrete Structures–2017*, Design
Japan Society of Civil Engineers, *Standard Specifications for Concrete Structures–2017*, Material & Construction
Japan Society of Civil Engineers, *Introduction to Mathematical Models of Material Characteristics*, Terminology of Constitutive Laws
K. Maruyama et al Japan Society of Civil Engineers, Concrete Library 101,*Recommendations for repair and strengthening of concrete strucrures with use of FRP sheets*
Japan Society of Civil Engineers, Concrete Library 69, *Research Report on the Mechanical Properties of Concrete*
K. Nishikawa, K. Uchida, A. Hiromatsu, K. Miyazaki, *Public Works Research Institute of Ministry of Construction, Cooperative Research on the Repair and Reinforcement of Concrete Members (I)–Research on the Reinforcement Effect of Carbon Fiber Sheet*
BSI, *Recommendations for Non-destructive Methods of Test for Concrete*, BSA 4408, Part5

BIBLIOGRAPHY

Chapter 1

Z.P. Bazant, B.H. Oh, Crack band theory for fracture of concrete. *Materials and Structures*, Vol.16, No.93, p.p.155–177

H. Yoshikawa, T. Tanabe, Analytic study on tensile stiffness of concrete members, *Journal of Japan Society of Civil Engineers*, Vol.366, No.V–4, pp.93–102

Hillerborg, Analysis of Crack Formation and Crack Growth in Concrete by Means of Fracture Mechanics and Finite Elements, *Cement and Concrete Research*, Vol.6, pp.773–782

D. Krajcinovic, Damage Mechanics: Accomplishments, Trends and Needs, *International Journal of Solids and Structures*, Vol.37, pp.267–277

J. MAZARS, A Description of Micro–and Macroscale Damage of Concrete Structures, *Engineering Fracture Mechanics*, Vol.25, Nos.5/6, pp.729–73

Q. Li, F. Ansari, Mechanics of Damage and Constitutive Relationships for High–Strength Concrete in Triaxial Compression, *Journal of Engineering Mechanics*, Vol.125, pp.1–10

W. Suaris, C. Ouyang, V.M. Fernando, Dmage Model for Cyclic Loadind of Concrete, *Journal of Engineering Mechanics*, Vol.116, No.5, pp.1020–1035

S. Yazdani, H.L. Schreyer, Combined Plasticity and Damage Mechanics Model for Plain Concrete, *Journal of Engineering Mechanics*, Vol.116, No.7, pp.1435–1450

E. Papa, A Damage Model for Concrete Subjected to Fatigue Loading, *European Journal of Mechanics, A/Solids*, Vol.12, No.3, pp.429–440

A. Pandolfi, A. Taliercio, Bounding Surface Models Applied to Fatigue of Plain Concrete, *Journal of Engineering Mechanics*, Vol.124, No.5, pp.556–564

K. Al-Gadhib, M.H. Asad-ur-Rahman Baluch, CDM Based Finite Code for Concrete in 3-D, *Computers and Structures*, Vol.67, pp.451–462

Al-Gadhib, M.H. Baluch, Damage Model for Monotonic and Fatigue Response of High Strength Concrete, *International Journal of Damage Mechanics*, Vol. 9, pp.57–78

J.F. Lopez, Simplified Model of Unilateral Damage for RC Frames, *Journal of Structural Engineering*, Vol.121, pp.1765–1772

J.F. Lopez, Frame analysis and continuum damage mechanics, *European Journal of Mechanics, A/Solids*, Vol.17, No.2, pp.269–283

Chapter 2

H. Kupher, H.K. Hilsdorf, H. Rusch, Behavior of Concrete Under Biaxial Stress, *Journal of the American Concrete Institute*, Vol.66, No.8, pp.656–666

N.S. Ottosen, A Failure Criterion for Concrete, *Journal of the Engineering Mechanics Division*, Vol.EM4, pp.527–535

J. Skrzypek, A. Ganczarski, *Modeling of Material Damage and Failure of Structures Theory and Applications*, Springer, pp.3–17

K. Maekawa, J. Takemura, M. Irie, Tri-axial Stress Effect on the Nonlinear Plasticity and Fracture of Concrete, *Proceedings of the Japan Concrete Institute*, 11–1, pp.253–258

I. Imran, S.J. Pantazopoulou, Experimental Study of Plain Concrete under Triaxial Stress, America Concrete Institute, *Materials Journal*, Vol.93-M67, pp.589–601

A. Benellal, R. Billardon, J. Lemaitre, Continuum Damage Mechanics and Local Approach to Fracture, Numerical Procedures, *Computer Methods in Applied Mechanics and Engineering*, pp.141–155

Y. Toi, J.G. Lee, H. Ioku, Element-Size Independent Elasto-Plastic Damage Analysis of Framed Structures, *Transaction of the Japan Society of Mechanical Engineers, Series A*, Vol.67, No.653, pp.8–15

Japan Society of Civil Engineers, *Standard Specifications for Concrete Structures–2002*, Structural Performance Verification, pp.26–28

J. Lemaitre, *A Course on Damage Mechanics*, Springer, pp.12–14

S. Murakami, *Continuum Damage Mechanics*, Morikita Shuppan Co., Inc., pp.9–10

Chapter 3

S. Murakami, Damage Mechanics and its Application to Fracture Analysis, *The Japan Society for Industrial and Applied Mathematics*, Vol.5, No.4, p.p.329–345

K. Kamiya, S. Murakami, Formulation of Damage Evolution Equation for Elastic-Plastic-Damage Materials in Stress Space, *The Society of Materials Science, Japan*, Vol.45, No.8, pp.893–900

Y. Toi, S. Hirose, Identification and Prediction of Mechanical Properties of Metals by Damage Mechanics Models, *Transactions of the Japan Society of Mechanical Engineers*, Vol.69, No.679, A, pp.530–537

H. Tanaka, Y.Toi, K. Maeda, T. Sakai, Damage and Failure Analysis of Brittle Structural Elements Reinforced by Carbon Fiber Sheet, *Transactions of the Japan Society of Mechanical Engineers*, Vol.72, No.716, A, pp.405–411

Chapter 4

F.J. Vecchio, M.P. Collins, Predicting the Response of Reinforced Concrete Beams Subjected to Shear Using Modified Compression Field Theory, *America Concrete Institute, Strututral Journal*, pp.258–268

H.T. Hu, W.C. Schnobrich, Nonlinear Finite Element Analysis of Reinforced Concrete Plates and Shells Under Monotonic Loading, *Computers and Structures*, Vol.38, No.5, pp.637–651

M.A. Polak, F.J. Vecchio, Nonlinear Analysis of Reinforced-Concrete Shells, *Journal of Structural Engineering, America Society of Civil Engineers*, Vol.119, No.12, pp.3439–3462

T.H. Kim, Nonlinear Analysis of Reinforced Concrete Shells Using Layered Elements with Drilling Degree of Freedom, America Concrete Institute, *Structural Journal*, Vol. 99, pp.418–426

B. Taljsten, Strengthening of Beam by Plate Bonding, *Journal of Materials in Civil Engineering*, Vol. 9, pp.206–212

M. Zako, T. Tsujikawa, Development of Computer Program for Fracture Simulation of Composite Structures, *The Society of Materials and Science, Japan*, Vol.42, No.474, pp.250–254

H. Niu, Z. Wu, T. Asakura, A Numerical Analysis on Bonding Mechanism of FRP-strengthened Concrete Structures Using Nonlinear Fracture Mechanics, *Proceedings of the Japan Concrete Institute*, Vol.21, No.3, pp.73–79

Z. Wu, H. Niu, Study on Debonding Failure Load of RC Beams Strengthened with FRP Sheets, *Journal of Structural Engineering*, Vol.46, pp.1431–1441

H. Yoshizawa, Z. Wu, H. Yuan, T. Kanakubo, STUDY On FRP–Concrete Interface Bond Performance, *Journal of JSCE*, Vol.662, No.V–49, pp.105–119

H. Morikawa, T. Kamotani, H. Kajita, Evaluation of Delamination Characteristics on Rc Beams Bonded with CFRP Sheet and Application to Smeared Crack Type Fem Analysis, *Journal of JSCE*, Vol.802, No.V–69, pp.15–31

T. Sugitama, A. Kobayashi, M. Saitoh, Study on Durability of Carbon Fiber Reinforced Plastics, *Proceedings of Symposium on Application of Advanced Reinforcing Materials to Concrete Structures*, Japan Concrete Institute Hokkaido Chapter, pp.49–56

L. Jeeho, Plastic–Damage Model for Cyclic Loading of Concrete Structures, *Journal of Engineering Mechanics*, Vol.124, No.8, pp.892–900

S. Deng, Y. Lin, Influence of Fiber–matrix Adhesion on Mechanical Properties of Graphite/Epoxy Composites, I. Tensile, Flexure and Fatigue Properties, *Journal of Reinforced Plastics and Composites*, Vol.18, pp.1021–1040

H. Tanaka, Y. Toi, K. Maeda, T. Sakai, Fatigue Fracture Analysis of Concrete Structures by Damage Mechanics Model, *Proceedings of the Computational Engineering*, Vol.10, pp.555–558

H. Tanaka, Y. Toi, *Fatigue Fracture Simulation of Concrete Element Strengthened with Carbon Fiber Sheet*, Japan Society of Simulation Technology 25th, pp.149–152

H. Tanaka, Y. Toi, Adhesive Failure Analysis of Structural Elements Reinforced with Carbon Fiber Sheets, *Transactions of the Japan Society of Mechanical Engineers*, Vol.72, No.724, A, pp.2007–2014

H. Tanaka, Y. Toi, K. Maeda, T. Sakai, Damage and Fracture Analysis of Brittle Structural Elements Reinforced with Carbon Fiber Sheets, *Journal of Environment and Engineering*, Vol.3, No.1, pp.111–121

H. Tanaka, H. Yanagihara, N. Sasaki, Adhesive Strength of Injectable Inorganic Anchor Material in Concrete, 9th International Conference on Geotechnique, Construction Materials and Environment, 9108

H. Tanaka, Adhesive Fracture Modes Between Steel Anchor and Injectable Material and Concrete in Consideration of Wedge–Shaped Effect, 11th International Conference on Geotechnique, Construction Materials and Environment, gxi219

Japan Concrete Institute, Concrete Diagnosis and Maintenance Technology '09 Basic Version, pp.210–214

Chapter 5

Y. Toi, The Finite Element Method and the Methods of Computational Discontinuum Mechanics, *The Japan Society for Industrial and Applied Mathematics*, Vol.3, No.4, pp.275–291

M. Asai, K. Terada, An Analysis of Propagating Discontinuities by the Finite Cover Method, *Journal of Applied Mechanics, Journal of Japan Society of Civil Engineers*, Vol.6, pp.193–200

Y. Toi, S.S. Kang, Mesoscopic Natural Element Analysis of Elastic Moduli, Yield Stress and Fracture of Solids Containing a Number of Voids, *International Journal of Plasticity*, Elsevier, pp.2277–2296

K. Matsumoto, Y. Sato, T. Ueda, Fracture Mechanism and Prediction of Deformation of Mortar Under Time–Dependent Loads by Meso–Scale Analysis, *Journal of Japan Society of Civil Engineers (E)*, Japan Society of Civil Engineers, Vol.66, No.4, pp.380–398

Y. Shintaku, K. Terada, *Cohesive-force Embedded Damage Model and Its Application to Crack Propagation Analyses, Transaction of the Japan Society for Computational Engineering and Science*, Paper No.20160011

Japan Society of Civil Engineers, *Standard Specifications for Concrete Structures–2018*, Maintenance, pp.70–87

Japan Concrete Institute, *Concrete Diagnosis and Maintenance Technology '09*, Basic Version, pp.181–199

Japan Concrete Institute, *Fundamentals of Concrete Technology '09*, pp.210–212

K. Kobayashi, T. Miyagawa, H. Morikawa, S. Igarashi, T. Yamamoto, T. Miki, *Concrete Structure the 5th*, Morikita Shuppan Co., Inc., pp.35–38

K. Togawa, H. Okamoto, H. Ito, T. Toyofuku, Y. Mitsuiwa, K. Yokoi, Y. Aoki, N. Takeda, *Concrete Structural Engineering, the 5th*, Morikita Shuppan Co., Inc., pp.35–67

Index

A

ABAQUS, 1
Adhesion fracture, 45, 60, 61, 80
ADINA, 1
Ageing deterioration, 89, 92
Arc length method, 28
Axisymmetric problem, 80

B

Bond-splitting, 74, 75, 77, 79
Brittle failure, 87, 89, 93

C

Carbonation, 2, 89, 92
Cone-shaped fracture, 75, 79
Continuum damage mechanics, 2, 7, 10, 13, 35
Corrosion, 3, 92, 93
Crack elements, 27–30, 36, 42, 43
Critical damage value, 24, 28

D

Damage potential, 12, 14
Discontinuous mechanics, 94
Drucker-Prager, 14, 24, 62, 63, 68

E

Effective equivalent stresses, 14, 35, 65
Effective stresses, 7, 13, 15, 16
Elasto-plastic damage
constitutive equations, 7, 13, 16, 17, 19, 27
laws, 4, 13, 18, 19, 29, 55, 61
tensor, 12, 20, 80
Energy equivalence principle, 13
Equivalent plastic strains, 14–16, 35, 38, 41, 80, 86, 87
stresses, 80–83, 86–88

F

Fatigue adhesion fracture, 45, 61
Finite element method, 1, 2, 4, 7, 12, 17, 27, 55, 80, 91
First invariant of stress, 14, 62
Fracture mode, 4, 47, 53, 60, 61, 74, 75, 77–80, 90
Freeze-thaw, 69–71
Fully Coupled Analysis, 4, 36–39, 41–43, 58

H

Homogenization method, 85

I

Inclination angle, 76–80
Increments of
elastic damage variables, 35
plastic damage variables, 35
Interfacial detached fracture, 83

K

Kachanov, 2

L

Lemaitre, 2, 12, 14
Limit state design method, 91
Locally Coupled Analysis, 4, 36–39, 57, 59, 60

M

Maximum principal stress, 62, 63, 68
Mesoregion, 91, 92
Miner's law, 10
Modified Newton-Raphson method, 28
Murakami, 2

N

NASTRAN, 1

P

Perturbation method, 28
Plane strain element, 29, 56
Plastic-hardened parameters, 14, 15
Plastic potential, 13–15
Post-installed anchor, 45, 76
Pre-crack, 86
Pre-fatigue, 45, 50–55, 57–60
Proportional coefficients, 15
Pull-off test, 52–54

R

Representative volume element, 5, 8, 11, 89–92, 94
Return mapping, 80, 86

S

Salt damage, 2, 89, 92
Second invariant of stress deviator, 14, 63
Serviceability limit state, 89, 91
Shear strength, 7, 45, 48, 86, 87
Smeared crack, 94
Softening region, 10, 19, 21, 28, 80
Strain energy release rate, 12, 14, 16, 35
Strain equivalence principle, 7, 13, 14
Suaris, 2

T

Tangential stiffness tensor, 29, 80
Thermal expansion coefficients, 5, 45, 68, 71–73, 81
Tresca, 62, 68, 88

U

Ultimate limit state, 89, 91